Paradigms in Cartography

Pablo Iván Azócar Fernández
Manfred Ferdinand Buchroithner

Paradigms in Cartography

An Epistemological Review of the 20th and 21st Centuries

Springer

Pablo Iván Azócar Fernández
Departamento de Cartografía
Universidad Tecnológica Metropolitana
Santiago de Chile
Chile

Manfred Ferdinand Buchroithner
Institute for Cartography
Dresden University of Technology
Dresden
Germany

ISBN 978-3-642-38892-7 ISBN 978-3-642-38893-4 (eBook)
DOI 10.1007/978-3-642-38893-4
Springer Heidelberg New York Dordrecht London

Library of Congress Control Number: 2013940745

Printed on acid-free paper

Springer is part of Springer Science+Business Media (www.springer.com)

Preface

This book represents the outcome of a fruitful and amicable cooperation over the past years. It is the result of intensive and partly controversial discussions between representatives of two "scientific generations". One of them (Manfred Ferdinand Buchroithner—MFB), initially coming from hardcore geosciences *and* philosophy and subsequently, motivated by years in the international mapping business, professionally moving to cartography; the other (Pablo Azócar Fernández—PAF), starting with geography/cartography and then developing an ever-increasing interest in theoretical and philosophical aspects of cartography. Influenced by his academic teachers in philosophy Rudolf Freundlich, Ernst Topitsch and Rudolf Haller MFB grew up in the tradition of the Vienna Circle and the thinking of Karl Popper and Bertrand Russell. In the early stages of his cartographic activities he was influenced by the epistemological views of his teacher at ITC (Enschede, NL) Sikke A. Hempenius and his predecessor at the Chair for Cartography at TU Dresden, Germany, the great East German theoretician Rudi Ogrissek. PAF was molded by the Latin-American cartographic thinking. Lectures about *Geography's Theory* at the beginning of 1990s, taught by the distinguished teacher Margarita Riffo Rosas (University of Chile), permitted him to immerse into philosophical topics related to geography. Soon, the question arose why these analyses are not applied to the field of cartography. After this motivational early phase of experience PAF's main focus turned to the relationships between philosophy, epistemology, cartography and mapping. During his doctoral research in Dresden, Germany, he deepened the aforementioned topics which are not sufficiently studied in South America.

The interest in the growing of our book by colleagues in Santiago and Dresden is gratefully acknowledged. We want to thank Wolf Günther Koch, Alexander Wolodtschenko and Dirk Burghardt, Dresden, for their continuous provision of the literature on the topics studied. Particular thanks go to Török Zsolt, Budapest and W. G. Koch for their critical reading of various stages of the book and their indispensible comments. We are also grateful to Mrs. Sharma and the staff of the TU Dresden Institute for Cartography. Pilar Correa from the Metropolitan Technological University (Universidad Tecnológica Metropolitana—UTEM, Chile) took care of the finalisation and homogenisation of the figures which we two authors created with our limited skills in graphic software. Our sincere appreciation has also to be extended to UTEM for the financial support during PAF's studies in Germany. Dalia Varanka (Rolla, Missouri, USA) undertook the invaluable task to bring our typescript into English which is also legible for native speakers. With her profound background in both cartography and philosophy as well as her experience in editorial issues she certainly improved the value of this book significantly. Thank you, Dalia.

Without the uncomplicated cooperation of Christian Witschel and Agata Oelschläger from Springer Verlag, Heidelberg, this book would not have been ready at the time of its publication. Our gratitude is also conveyed to them. Finally, we wish to thank our families for the time they had to abstain from our company because we were dealing with another love of ours, the epistemology of cartography.

During the writing of this volume, controversial views set forth by various authors had, after sometimes controversial discussions between the two of us, to be synthesised. The authors are aware that this book will certainly also raise contradictions among the readership. If this can be achieved and a scientific discourse can be triggered, one of the intentions of this work has already been reached. Cartography needs more research in the field of epistemology! It should, however, not be concealed, that over the years of manuscript generation also the authors' views of things underwent some modifications—sometimes even with significant changes in direction (which can, in part, be traced by their publications). Science is dynamic! And in this sense the current book is not only intended to be a historical-epistemological investigation and documentation of twentieth and twenty-first century cartography but also a triggering agent for further research in this field. If this can be achieved, the authors have reached a good part of their intentions.

Santiago de Chile | Pablo Iván Azócar Fernández
Dresden, May 2013 | Manfred Ferdinand Buchroithner

Contents

Introduction

The first mention of cartography as a *science* occurring in an internationally binding official document occurs in a report issued by the United Nations Department of Social Affairs in 1949 (United Nations 1949) entitled "Modern Cartography. Base Maps for World Needs". In their Introduction the involved experts state that:

> Modern cartography covers a wide and complex range of subjects. Maps may portray political boundaries or oceanic depth, the geological structure of the earth's surface or the density and distribution of its population; they may be used to represent, in graphic form, inventories of the world's natural and industrial resources or of educational facilities. Military requirements, it is true, still retain a high priority in map making. However, it is equally true that the services provided by cartography are being increasingly utilized for peaceful purposes. This fact is perhaps the dominant feature of the development of cartography during the last hundred years. The progress of the *Science of Cartography* [italics not in original document] is of equal concern to all nations.... (United Nations 1949: Introduction).

Nevertheless, cartography has commonly been considered to be both *science* and *technique*, as well as an *art* in the design, construction and study of maps. An analysis of cartography in *philosophical* and *epistemological* terms, however, raises important aspects. A *positivist* perspective of sciences, for instance, denies art as a part of cartography. On the other hand, from a *humanist* point of view the subjective aspect of maps is emphasised, and at the same time the scientific view on the discipline is criticised. If one focuses only on the *technological* aspect of cartography, emphases are put on a *pragmatic* vision of the "reality", leaving aside all other aspects of map analysis. In turn, if maps are examined from *historical* and *hermeneutic* points of view, then they are considered *texts* which convey a *political–cultural–social context*, in which power relation and subjectivity acquire relevance. Therefore, in this book, besides highlighting epistemological and philosophical issues, cartography is considered a solid body of knowledge to understand our world with all its different facets. The authors try to follow an *analytical and holistic view* of the more recent historical developments in cartography. This is also corroborated by not only treating map theory but also map-use aspects and can thus demonstrate the *theoretical eclecticism* in cartography.

In 1982, in a precursor to his famous book about the theory of cartography, published 5 years later, Rudi Ogrissek (4.9.1926–27.9.1999) preempts several of his later thoughts and displays, in particular, some seminal illustrations (Ogrissek 1982). This 80-page booklet, which is to a great deal based on findings of Eastern, especially Russian, cartographers, and which was primarily meant as a textbook for East German students, is remarkable for various reasons: Among others, Ogrissek therein coined the excellent term '*imagination map*' ('*Vorstellungskarte*'), thus drawing on models by A. M. Berljant, K. A. Salishchev and A. S. Vasmut (Ogrissek (1982) and thus being in a sort of contrast to the early papers by Gould and White (1974) and Kishimoto (1975) who use the nowadays widely accepted term "mental map". Anyway, this work represents one of the first more comprehensive publications on the theory of cartography outside the region of Slavic languages and brought, for the first time, ideas of the former Soviet cartographers into a non-Russian language.

Already in 1986, Joel Morrison, co-author of the renowned textbook on cartography (Robinson, Sale and Morrison 1978) and president of the International Cartographic Association (ICA) from 1984 to 1987 (Wolodtschenko 2008) criticised the missing philosophical research in cartography (Morrison 1986). Consequently, this led to the establishment of an initial working group and later a Commission on Theoretical Cartography within ICA. At the 16th International Cartographic Conference 1993 in Cologne, Germany, a first report of this international working group was presented, the results of which have later been published (cf. Kanakubo et al. 1993). This paper nicely shows how scientific paradigms work (hint by courtesy of Zsolt Török, Budapest, written communication 2012).

The present book incorporates Thomas Kuhn's *Concept of Paradigm*. It deals with the different thoughts and tendencies which contemporary cartography has experienced during the so-called modern and postmodern periods. Thus, current trends in cartography are analysed regarding the extent to which they can be identified as paradigm shifts. For each cartographic trend it is necessary to keep in mind the theoretical schemes of a scientific discipline (study object, research aim, method and results) which are supposed to determine its body of knowledge. The above will reveal if cartography has been having its own paradigms, which would imply it has its own autonomy, or if its body of knowledge comes from other sciences. Finally, the theoretical character of the study should help us to understand the discipline beyond its practical-technological aspects. It represents a theoretical contribution, because the analysis of the cartographic tendencies stresses, from an epistemological viewpoint, their scientific, deconstructivist and ontological levels.

Regarding the need for theory, this book is not only supposed to be a contribution to theoretical cartography but also to represent a sort of stimulating contribution to the development Modern cartography covers a wide and complex range of subjects. Maps may portray political boundaries or oceanic depth, the geological structure of the earth's surface or the density and distribution of its population; they may be used to represent, in graphic form, inventories of the world's natural and industrial resources or of educational facilities. Military requirements, it is true, still retain a high priority in map making. However, it is equally true that the services provided by cartography are being increasingly utilized for peaceful purposes. This fact is perhaps the dominant feature of the development of cartography during the last hundred years. The progress of the *Science of Cartography* [italics not in original document] is of equal concern to all nations.... (United Nations 1949: Introduction) of cartography in the twenty-first century. We feel that currently a growing awareness for the need of a theoretical basis of our passion, cartography, is arising among both academics and professional cartographers. It seems that today this need is no longer questioned.

Whenever possible, the authors based their statements on the primary literature. In some rare cases, however, this was not feasible. We are, albeit, well aware that the secondary literature sometimes might include disputable and/or superficial statements.

The book is divided into nine chapters and a Reference section. Chapters 1 and 2 analyse the theoretical bases. The link among philosophy—emphasizing the relationship between object, subject and image—epistemology and cartography is analysed. In Chap. 1, in the traditional way of knowledge theory, first the different so-called "*isms*" are examined in their relationship between subject and object. Second, three epistemological-philosophical perspectives are analysed: *positivism-empiricism*, *neopositivism* (logical positivism) and *postmodernism* (poststructuralism). Within every period the impact on cartography and mapping is described. Then, the consolidation of geography as a scientific discipline and its effect on cartography during the positivist period is considered. Cartography as a science or discipline has then been taking the scientific features corresponding to the positivist context.

In Chap. 2 three great philosophers are treated. First, the traditional legacy of Immanuel Kant and then two contemporary philosophers of the logical positivism: Ludwig Wittgenstein and Karl Popper. Emphases are put on the question of how cartography grasps epistemological aspects in knowledge construction. Although the two aforementioned authors did not directly write about cartography and mapping, their legacy has had an important impact on our understanding of maps. The evolution of contemporary cartography can be linked to the development of Wittgenstein's thinking (Wittgenstein's *First and Second Philosophy*). On the other hand, the cartographic products and the different stages of map creation can be linked to Popper's Three Worlds Theory. The statements of these two philosophers are related under stances of scientists who rather belong to the cartographic field such as Herbert (2002), Lois (2009) and Cauvin et al. (2010). During postmodernism new tendencies and perspectives arose from *social theory*. Their relationships with cartography and mapping are also discussed. Also, this part of the book refers to the contribution of Immanuel Kant, especially his links with geography, and the concepts of space and time. The aim of this comprehensive Chap. 2 is to lay the theoretical bases of or rather for cartography, as this should help to understand the discipline beyond its technological issues.

For the remaining part of this book the major theoretical fundamentals are made up of the paradigm concept developed by Thomas Kuhn in the context of the history and philosophy of science (Kuhn 1962, 1970). This topic is treated in Chap. 3. The term *paradigm* has several interpretations, but in general includes a scientific community (in a particular field of knowledge) in which common aims and criteria during a determined period of time are shared. These periods are also called *normal science*. The replacement of a paradigm by another one is known as *scientific revolution* or *crisis period*. Therefore, our views incorporate Kuhn's epistemological concept of paradigm in order to be later applied to the discipline. Thomas Kuhn's ideas are taken as given and valid, they allow to analyse the extent to which contemporary tendencies in cartography can be identified as paradigmatic elements within the scientific community.

Chapters 4–6 focus on a review of the state of the art and recent trends in cartography. Chapter 4 deals with several authors and theoreticians who analysed the discipline during its contemporary development under certain paradigms and currents (Peterson 2002; Perkins 2003; Wood 2003; Ramirez 2004; Edney 2007; Ormeling 2007; Sui and Holt 2008; Cauvin et al. 2010). It also discusses geovisualisation (DiBiase 1990; DiBiase et al. 1992; MacEachren 1994, 1995; MacEachren and Kraak 2001), analytical cartography (Moellering 2000, 2001a, b; Tobler 1976, 1979) and cyber-cartography (Taylor 2005, 2009) approaches.

Beginning in the second half of the twentieth century, various authors, from Robinson (1952) up to MacEachren (1995) and Taylor (2005), established and ascertained scientific trends. Tendencies such as cartographic language, cartographic communication model, analytical cartography, geovisualisation and cyber-cartography are framed into a neopositivist approach of modernity. Therefore, in epistemological terms, cartography and mapping try to reach a representation (depiction) of the geographical space as veridically as possible: accurate, precise, secure and objective. This is considered the main aim of cartography: to reach an objective representation of the world. In other words, in this case, the metaphor of the map as *reflection* or *mirror* of "reality" is valid.

Although in Chap. 4 *critical cartography* is described as a new tendency, beginning in the 1980s, Chap. 5 still deepens its treatment in the postmodernist context. The critical perspective is a historicist view of cartography that poses a conception of the map as a text or vehicle of *power and knowledge* (Harley 1988a, b, 1989, 2001). Thus, there exists a historical critique of the power of maps in different times and places, and a contemporary critique of maps regarding ethical considerations and values (Crampton and Krygier 2006; Wood and Krygier 2009; Crampton 2010). The chapter examines John B. Harley's legacy and his link with postmodernist thinkers such as Michel Foucault and Jacques Derrida. In this way, the critical approach points out that maps act as *rhetorical* devices which implicitly and explicitly pass messages of hegemony and power in a specific social-political context. In this context, the map acts as a subjective device, biased, loaded with values and meanings. Thus, in agreement with the so-called postmodern authors, it is stated that the critical perspective constitutes a *paradigmatic shift* that breaks with the objectivity and neutrality claimed by the previous stances.

Continuing in the postmodern cartographic context, Chap. 6 focuses on new map conceptions which are challenging the previous ones (Latour 1987, 1999; Crampton 2003; Pickles 2004; Casti 2005; Wood and Fels 2008; Della Dora 2009). Additionally, since the end of the 2000s, some authors such as Rob Kitchin, Chris Perkins and Martin Dodge have taken up a *post-representational* attitude in which cartography and maps are seen to be beyond the previously established formal and positivist aspects. Consequently, the traditional *ontological* conception of maps is criticised and replaced by an *ontogenesis* conception (Kitchin and Dodge 2007; Kitchin 2008; Kitchin et al. 2009). This conception proposes that a map is not an epistemologically *stable and secure* product (as taken for granted in the scientific and critical approach), but rather the result of the moment: it is subject to continuous *re-creation* and *re-interpretation* according to the context in which it is situated. The map is seen *to be in action*. It is in a *state of becoming*. In this book this new perspective is discussed as a possible, ongoing *emergent paradigm* in Kuhnian terms, subject to considerations regarding the epistemological and ontological bases of the discipline.

In Chaps. 7 and 8 a methodological proposal to identify paradigms in cartography is set forth. Chapter 7 analyses whether at all there exists a possibility for paradigms in cartography as defined by Kuhnian terminology. Here, in methodological terms, two approaches are proposed. The first one is called "*criteria of contrast*" (e.g. study object, research aims, method and approach, results, etc.). These criteria are applied to identify formal and factual sciences and also to differentiate between regional and quantitative geography. The second methodological procedure is named "*tendency distribution in the epistemological space*". In this context, contemporary cartographic tendencies are located under the three philosophical-epistemological bases of modernity and postmodernity: positivism-empiricism, realism-structuralism, idealism-hermeneutics. As shown in Chaps. 4–6 several cartographic trends have occurred since the second half of the twentieth century. Will, in epistemological terms, these tendencies shape groups or clusters or will they remain isolated? If they were disparate trends, would we be able to propose some internal paradigm shifts within cartography? The two applied methodological criteria allow the identification of some internal *worldviews*—as termed by Kuhn—during the contemporary development of the discipline.

Chapter 8 discusses the results obtained in Chap. 7. At first, seven paradigm tendencies in cartography are proposed, based on the criteria of contrast and the opinions published by the authors that have been critically reviewed in the previous sections. On the one hand, the body of knowledge of the discipline is characterised through the distinction between the scientific and critical approaches (i.e. by contrasting paradigms), and on the other hand by the transition between both stances. Also, post-representational cartography is considered a paradigmatic proposal

Fig. 1 What this book is all about in Cartography. Figure adapted from http://www.helenleeauthors.com/2011/09/understanding-your-vocation-life/; http://streamlinetosucceed.com/wp-context/blogs.dir/50/files/2011/03/who-what-where-280.png

which challenges previous approaches. Here, the ontologically secure map is doubted. In Kuhnian terms this development deserves to be observed in an alert manner regarding a new worldview in cartography and mapping.

The second part of Chap. 8 returns to Kuhn. His scientific revolution theory is discussed within the scope of intrinsic paradigms in cartography. In epistemological terms, three levels are examined: scientific, sociological and ontological.

As a result of the last chapter the authors state that, if the development of cartography and mapping is considered to take into account the epistemological coordinates, then three paradigm shifts in Kuhnian terms can be postulated: scientific-empirical, critical and post-representational. Despite the fact that these three paradigmatic shifts have been triggered by technological development, the theoretical statements made in this volume go beyond technological issues in cartography and mapping. This theoretical posit is corroborated by philosophical and epistemological considerations about the development of the cartographic science. Finally, a chapter with conclusions and a comprehensive list with literature references are presented (Fig. 1).

Chapter 1
Philosophy, Epistemology, and Cartography

1.1 Introduction

At the outset, before all further statements in this book, we link philosophy, emphasizing the relationship between object, subject, and image, and epistemology, as the theory of scientific knowledge and contemporary cartography. Three epistemological perspectives are analysed: positivism and empiricism, neopositivism (logical positivism), and postmodernism (poststructuralism). The impacts of each of these periods on cartography and mapping are described. Initially, the consolidation of geography as a scientific discipline and its impact on cartography during the positivistic period is considered. Cartography as a discipline covers all the scientific features corresponding to positivism. Subsequently, new tendencies and perspectives arising from social theory during the postmodern period, and their relationship with cartography and mapping are also discussed.

1.2 Philosophy and Epistemology: Some Scopes

There exist several definitions of philosophy. According to Johannes Hessen a look at the evolution of philosophical thinking shows two elements that are essential for the concept of philosophy: (a) a conception of the individual person and (b) a conception of the universe. So, philosophy represents both *conceptions* simultaneously (Hessen 1976). As Johannes Hessen writes, 'Philosophy is an attempt of the human spirit to come to a conception about the universe by means of self-reflexion about its theoretical and practical functions' (cited after Vargas Mendoza 2006).[1]

To achieve this aim, philosophy attempts to formulate and answer some questions. Philosophical questions are fundamental. They are of two sorts, as mentioned above, *internal conception* (about me) and *external conception* (about the universe), implying that they are placed within the limits of human

[1] 'La filosofía es un intento del espíritu humano para llegar a una concepción del universo mediante la autorreflexión sobre sus funciones valorativas teóricas y prácticas.'

P. I. Azócar Fernández and M. F. Buchroithner, *Paradigms in Cartography*,
DOI: 10.1007/978-3-642-38893-4_1,

comprehension. This, again, means that philosophical questioning by the human spirit leads to both man's *interior world* and man's *exterior world*.

As part of philosophy, *epistemology* concerns the nature, origins, and limits of knowledge, i.e. a theory of the knowledge. The *philosophy of science* and the *epistemology of science* both concern *scientific knowledge*. According to Mario Bunge, epistemology, or the philosophy of science, is the branch of philosophy that studies scientific research and its product, scientific knowledge (Bunge 1998). Thus, theories of scientific knowledge differs from that of other types of knowledge, such as technical, technological, artistic, and religious knowledge. Scientific knowledge may be rational, factual, objective, methodical, self-corrective, or progressive, general, systematic, and accumulative.

On the other hand, since twentieth-century mapping and cartography have been considered the *science**, *technique* and *art* of the design, construction and study of maps, three important aspects regarding maps are implied: to imagine, to use, and to interpret them. When cartography and mapping are considered from a philosophical and epistemological point of view, then important considerations concerning our knowledge of the world arise (cf. i.a. Reder 2012).

1.3 Cartography and Knowledge of the World: Philosophical and Epistemological Implications

If cartography and mapping have contributed to the development of mankind's body of knowledge, then the way to conceive, to imagine or to understand a map according to a socio-political and cultural context varies. It is necessary to bear in mind that such contexts underpin the importance of a philosophical basis for understanding the physical world, and thus cartography can be assumed like contribution to the knowledge.

In his prominent book *Teoria del Conocimiento* [*Theory of Knowledge* 1976] Johannes Hessen states that there exist some partial problems regarding the General Theory of Knowledge such as possibility, origin, essence, and kinds of knowledge. Philosophically, three elements participate in all issues of knowledge: a *subject*, an *object* and an *image* (of an object which is captured by a subject). Different philosophical currents originated depending on the degrees of importance assigned to each of these elements, for instance *objectivism*, *subjectivism*, *realism*, *rationalism*, etc.

If knowledge is considered as the relationship between an *object* and a *subject* (having mutual contact), the question is: Can the subject really capture the object? In response to this question, the following perspectives arise: *dogmatism, scepticism, subjectivism, relativism, pragmatism* and *criticism*. When the dual structure of the cognoscible subject is considered to comprehend *rational knowledge* and *sensible knowledge*, the query is: Is reason or is experience the source and basis of human knowledge? This question leads to *rationalism, empiricism, intellectualism* and *apriorism*. When the relationship between the subject and the object is considered, the question arises: Is it really the object that determines the subject or is it the subject that determines the object? This question leads to the rise of *objectivism, subjectivism, realism, idealism* and *phenomenalism*. When the form or kind of knowledge is considered, the question is: Do other kinds of knowledge exist besides *rational knowledge*, e.g. *intuitive knowledge* or *experiential knowledge*? Last but not least, the *truth criterion* plays a role; that is asking if a particular type of knowledge is true. Here the question is: Which *criterion* indicate whether a specific *knowledge* is true or not? (For more details see Hessen 1976.)

All these epistemological problems regarding the theory of knowledge are also important when we think about maps, because maps have always been conceived as a form of knowledge or as devices which contribute to the knowledge of the world.

These considerations are relevant when *the* controversial question rises in cartographic literature: Do maps represent or do they create reality?[2] For this inquiry, terms like geographical space, Earth surface, terrain, landscape, and so on, are replaced by the *concept of reality*. The two verbal terms involved in the query (i.e. to represent, to create) give rise to different epistemological viewpoints. On the one hand, *to represent reality* implies elements of the *exterior world*, and in this case, notions about the object, objectivism, and empiricism (sensible knowledge) are engaged. On the other hand, *to create reality* entails an *internal* and a *mental world* (mind). Thus, conceptions about the *subject*, *subjectivism*, *rationalism* (*rational knowledge*), and *idealism* are involved.

Therefore, the answer to the previous question has philosophical and epistemological scopes depending on the respective point of view. When the map *represents reality*, it leads to the *scientific* approach of knowing the world, and when the map *creates reality*, it acts as a guide to the *humanistic* approach of knowing reality. Sometimes the map has the two functions at the same time. At some point there has to be an epistemological division because both approaches require different study objects, methodologies, and results.

Another significant consideration concerns the nature of the cartographic representation[3] of the relation between object, subject, and image. It is important to evaluate whether the a map corresponds to what it is supposed to *reflect* (from the external world) or to the memory of a subject that it *activates*. Again, the answer depends on the considered philosophical and epistemological perspectives. If a map *reflects* the world, then it leads to an *external* perspective (*objective*). On the other hand, if a map *actives* a subject's memory, then it leads to an *internal* perspective (*subjective*).

---

[2] This question is derived from an article by Laura Hebert (2002) postulating the entirety of both physical and human/social aspects.

[3] This scope is derived from Carla Lois (2009).

1.4 Different 'Isms' and Their Repercussions in Cartography and Mapping

In philosophy and the epistemology of science, it is common to classify trends of thought with the suffix 'ism'. Thus, different scientific conceptions such as *positivism*, *falsificationism*, *historicism*, *structuralism*, *relativism*, *objectivism*, *pragmatism*, *realism*, and others have been established. These tendencies can be analysed by identifying the visions and perspectives that distinguish them. Nevertheless, in this work, only some of the main tendencies that occurred during the modern and postmodern periods will be considered.

1.4.1 Positivism and Empiricism

During the eighteenth and nineteenth centuries in western societies, the following motto prevailed: Faith in progress, dominance of nature, and use of reason. Progress referred to the industrial and transportation revolutions, nature to the territories and landscapes, and reason was emphasised with respect to scientific issues. This conception was called Philosophical Positivism, introduced by Auguste Comte with his *Discours sur l'Esprit positif* (1844) [*A General View of Positivism* 1865]. The main exponents of this school are John Stuart Mill, David Hume, Bertrand Russell, Alfred Jules Ayer, and Rudolf Carnap.

According to George Wright, *positivism* shows three essential features: (a) methodological monism; (b) the assumption that exact natural sciences establish the ideal canon or methodology of all sciences, including the human and social sciences; and (c) the concept of causal explanation (Wright 1971).

Positivism considers a *monistic conception* of reality. *Monism* states that reality is unique, and it has to be understood and explained by physical science in the first instance, and following that, the social sciences. The empirical-inductive method, or also called the *positive method* (from detail to general), was used as the study approach for deepening knowledge about the physical and social world. This method comprises two stages: first, *induction*[4] leads from facts to laws and theories; and second, these laws and theories are established by means of predictive *deductions*.[5]

The *positivistic period* is characterised by emphasizing scientific knowledge. Facts and phenomena have to be observed, measured, verified, and expressed in *laws*. In other words, facts and phenomena of the physical world must be scientifically explained. Science uses causal explanations with a deterministic cause-

[4] *Induction* is the process which leads from details to the general, using particular observed data to derive general laws.

[5] *Deduction* proceeds from the general to the particular, using a general law to explain particular cases.

effect relation.[6] It first adopts the mechanistic Newtonian Model in physical sciences and later applies it to the biological Darwinian Model in human and social sciences (Gómez Mendoza et al. 1988). Another feature of science is its objectivity: observations and explanations must be free of value assignments (*neutrality of sciences*).

In addition, *empiricism* is 'an epistemology in which it is assumed that learning, memory, and ideas are primarily derived from one's sensory experiences' (Slife and Williams 1995). In other words, according to empiricism, knowledge comes from experience, from evidence, and from direct contact with reality. So, *empiricism* can be associated with *positivism*, because both of them originate from the positivistic viewpoint that reality is observable by human sensory perception.

Given the positive spirit of the epoch (especially in the nineteenth century), in science in general and in geography and mapping in particular, all exponents and researchers shared the positivist epistemological approach. This implied that a high scientific level developed within the positivistic movement.

On the other hand, during the period of positivism the following events impacted the development of the geographic discipline in Europe (Gómez Mendoza et al. 1988; Ortega Varcárcel 2000), and also established the basis for cartography and mapping.

1. European space is organising itself for the exploration of the world, triggered by the need for raw materials, food, and work[7] (Vico and Bentancor 1999).

The link between cartography and political geography is documented by Matthew Edney in his work "Recent Trends in the History of Cartography". He refers to several authors who investigated the "colonial and imperial cartography" as well as "cartography and nationalism" during the modern period (cf. Edney 2007). Crampton emphasizes the facts that governmaps in the context of cartographic political economy (cf. Crampton 2010). Cartographic research in the nineteenth century was carried out by the military and therefore it heavily focused on surveying technology and military topography (Kanakubo 1990). Thus, the European exploration was supported by mapping and cartography whose products, i.e., the maps, are invested in a *scientific key*.

2. A new profession developed: the scientist. His mission is to discover the order which exists in the world, in particular in the natural world[8] (Vico and Bentancor 1999).

[6] *Causality* is the relationship between an event (the cause) and a second event (the effect) where the second event is understood as a consequence of the first one.

[7] 'El espacio europeo se organiza para la exploración del mundo por necesidad de materias primas, alimento y trabajo'.

[8] 'Se crea una nueva profesión, el científico. Su misión es descubrir el orden que existe en el mundo, en particular el mundo natural'.

According to this assignment, it is no coincidence that topographic cartography precedes the thematic. At the first stage, the *formal maps* which depict real objects of the world were constructed by mapmakers. These topographic maps mainly contained natural elements and visible artificial elements of the Earth's surface. In the second phase, abstract or ideal elements were depicted as thematic maps.[9] Methods and techniques were improved to reach higher levels of precision, accuracy, and objectivity in map representations. These cartographic methods were coined within the epistemological framework contemporarily in force.

The first independent, professional cartographers appeared in the nineteen century. According to Árpád Papp-Váry "cartographers of that century only became cartographers through their practical work; they originally were geographers, copperplate engravers, engineers, and army officers" (Papp-Váry 1989: 103).

3. Authors of geography are individuals of their time, influenced by changing ideas and collaborators in the expansion of knowledge[10] (Vico and Bentancor 1999).

The Germans Alexander von Humboldt (1769–1859) and Carl Ritter (1779–1859) were early strivers for cartographic depictions and the precursors of modern geography (Schmithüsen 1970; Abler et al. 1971). The former analysed observable relationships among physical phenomena, and the latter introduced the human factor into the man-environment relationship studies. Theses efforts resulted in systematic cartographic representations, especially with Humboldt's contributions in the field of physical geography (i.e. relief representation through contour lines). The circle around Carl Ritter and Hermann Berghaus (1828–1890) fine-tuned the concept of *Kartographie,* finally leading to the workclass map public Wilhelm Perthes in Gotha (courtesy written communication by Zsolt Török 2012; Hantzsch 1902).

Similarly, 'the geographer's attention was turned towards maps which influenced the progress in geographical research systems' (Kanakubo 1990: 4). Therefore, cartography and mapping are considered forms of scientific knowledge.

4. In the second half of the nineteenth century, formal education is institutionalised, explorations promoted, and research is systematised[11] (Vico and Bentancor 1999).

If cartography is conceived as a form a spatial representation, it is then not strange that the formalised study curricula for geographical topics also incorporated cartographic issues such as scale, projection, symbols, and colour, among

[9] The historical evolution of thematic cartography is i.a. described by Colette Cauvin et al. (2010).

[10] 'Los autores de Geografía (...) son hombres de su época, influidos por las ideas del momento y colaboradores en la expansión del conocimiento'.

[11] 'En la segunda mitad del siglo XIX se institucionaliza la enseñanza formal, se fomenta la exploración y se sistematiza la investigación'.

other. The construction and updating of atlases was a form of systematisation of acquired spatial and geographical knowledge.

5. The establishment of learned societies enables exploration of the world and academic discussion about acquired knowledge[12] (Vico and Bentancor 1999).

Concerning cartography and mapping, research in map projection, relief representation, map colour, and the construction of various types of atlases were efforts made during this period. Thus, the second half of the nineteenth century was the period during which *theoretical cartography* began to flourish. Geographical associations provided the basis for the development of cartographic associations during the second half of the twentieth century, e.g., the International Cartographic Association (ICA), founded in 1959.

Crampton (2010), quoting Turnbull (2003), points out that:

> A centralized knowledge base went hand in hand with the emerging modern political state, a unique system of measurement (the metric system), and common set of instruments [...]. Such a centralized, almost panoptic system is characteristic of rational scientific knowledge-creation (Crampton 2010: 56).

This statement illustrates the positive spirit of the great surveys of the eighteenth and nineteenth century, who did their job using the Enlightenment principles of rationality, precision, and instrumental surveying and mapping. In other words, cartography and mapping—as a form of knowledge—fulfilled the *epistemological objectives* of the positivistic period.

1.4.2 Logical Positivism or Neo-Positivism

In the first third of the twentieth century, a questioning and criticism of the conceptions of positivism evolved, especially regarding new developments in sciences. Consequently, a new proposal for the philosophy of knowledge was propelled. This epistemological reflection is known as *logical positivism*, especially in the scientific and philosophical context of the German-speaking region. A key feature of logical positivism or *logical empiricism* is the experimental verification of theoretical statements through a verification process and their validation. This new approach is a revitalised and changed formulation of the positivistic legacy.

In 1929, a group of distinguished scientists and philosophers of science linked with the universities of Berlin and Vienna, founded a collective known as the

---

[12] 'La creación de Sociedades Científicas habilita la exploración del mundo y la discusión académica de los conocimientos adquiridos.'

Vienna Circle ("Wiener Kreis")[13] (Murzi 2004). Core members of this group, also called the Ernst Mach Society (*Verein Ernst Mach*) in honour of the physicist and philosopher Ernst Mach were Moritz Schlick, Gustav Bergmann, Rudolf Carnap, Philipp Frank, Hans Hahn, Tscha Hung, Victor Kraft, Karl Menger, Richard von Mises, Marcel Natkin, Otto Neurath, Olga Hahn-Neurath, Theodor Radakoviv, Rose Rand, and Friedrich Waismann. The Vienna Circle's philosophers sought to re-conceptualise empiricism by means of their approach to interpretation in the physical and formal sciences. 'Their radically anti-metaphysical stance was supported by an empiricist criterion of meaning and a broadly logistic conception of mathematics. They denied that any principle or claim was synthetic *a prior*'[14] (Uebel 2006, digital text: no pagination).

This new philosophical perspective is also known as *neo-positivism*. This approach considers several aspects of the traditional positivism of the nineteenth century, sharing aspects such as: monist conception of reality, unity of sciences, and emphasis on the explanation[15] of facts and phenomena, more than mere comprehension.[16] But, there is an essential difference to the old positivist method: this time, the hypothetical-deductive method (from the general to the particular) is used. Induction is replaced by the deductive-via which goes from logical statements to the observation of facts. These facts turn into verifiers of the statements. The neo-positivist thinking has three pillars: *the fundamental role of the facts*; the introduction of *theoretical constructs,* and the use of the *formal language* (Ortega Valcárcel 2000).

In the scientific context, high objectivity—meaning the separation between facts and values—is required, as well as an extreme rigor concerning the accuracy of the results, verification of hypothesis, and validation of pertinent theories. Besides occupying the common method of sciences, the scientific method, scientists must use a common language that is accurate and precise. That would be the mathematical and logical languages, as influenced by the Vienna Circle. Regarding the specific function of language, José Ortega Valcárcel states:

---

[13] The Vienna Circle unites the physical and *sensorial empiricism* of Ernst Mach with the logical mathematic school of Bertrand Russell (1872–1970) and their adherent Ludwig Wittgenstein (1889–1951). This organisation created its own conception of philosophy of science and knowledge of positivistic character, including the rationalistic tradition. One may state that around World War II, with Vienna as a centre, a transition occurred from *constructive realism* to *critical rationalism* which is also, to some extent, reflected in the philosophical development of Karl Popper (1902–1994).

[14] A synthetic *a priori* proposition is one that being *a priori,* i.e., universal and necessary in character, has the extensive property of the *a posteriori* propositions, thus allowing to increase our knowledge.

[15] *Explanation* leads to understanding reality through laws and principles (knowledge of general cases).

[16] *Comprehension* leads to understanding reality through descriptive terms (knowledge of particular cases) without establishing laws. In epistemological terms, there exist no laws for uniqueness.

> Semiotics is the ultimate foundation of scientific communication, removing to the thinking as subjective activity, except in the strict work of combining the signs. In it the deductive or analytical process is based, whose tautological nature assures the quality of verifying[17] (original emphasis, translated from Ortega Valcárcel 2000: 201).

According to Ortega Valcárcel (2000), epistemologically there exist two worlds: the world of *analytic knowledge* and the world of *synthetic knowledge*. The former one is recognised as a rational activity, the world of logical statements; analysis in a strict sense, deduction, the world of signs and their rules, the realm of truth. The world of theoretical statements acquires an absolute pre-eminence; thus the new philosophy is *analytical*. The latter one—synthetic knowledge—is the world of experience, of facts, i.e., it is empirical. This synthetic knowledge verifies the validity of theoretical statements. Therefore, the truth or falsehood of scientific theories is verified.

In summary, the neo-positivistic approach has the following characteristics regarding the bonds among philosophy, epistemology and science: The new science is structured according to abstract parameters that attempt to explain reality on the basis of theoretical models. The new science has a formal structure which is expressed by a logical language. The hypothetical-deductive method is the unique scientific method capable of creating new knowledge. Thus, a new occupation has come into existence: the professional scientist who is contracted for doing prospective studies. Science finally acquired a practical character. Its new concern was created an order which has to adjust itself to the theoretical models.

1.4.3 Geography and Cartography in View of Logical Positivism

Both geography and cartography were impacted by the neo-positivistic approach. In the geographic discipline, the previous regional perspectives were criticised and new proposals arose. *Quantitative geography* was now the new paradigm[18] in geographical studies. Relationships and spatial distributions, in terms of spatial geometry, have been the study objects of geography under this approach. Any phenomenon (natural or human) can be analysed and spatially expressed in geometric form. The majority of these spatial relations were depicted through cartographic methodologies (Chorley and Hagget 1967).

On the other hand, the objective of quantitative geography is to explain spatial phenomena through their prediction. Thus, its results must be expressed by laws

[17] 'La semiótica es el fundamento último de la comunicación científica, desalojando al pensamiento como actividad subjetiva, salvo en la estricta labor de combinar los signos. En él reposa el proceso deductivo o analítico, cuya naturaleza tautológica le asegura la cualidad de *verdadero*'.

[18] The term *paradigm* introduced by Thomas Kuhn in 1962 will be discussed in the following section.

applying the hypothetical-deductive method. The tools used in this method are: models, hypotheses, laws, and theories (Harvey 1969). Nevertheless, a relevant aspect is the use of quantitative technologies such as mathematics, statistics, and cartographic depictions of correlations. In cartography, processes related to in- and output of geospatial data are automated. The intensive application of these technologies and methodologies led to a revolution in the analysis and synthesis of geographical studies.

Similarly, in the context of the neo-positivistic paradigm, authors like Richard Chorley, Peter Hagget, and David Harvey developed the theoretical-scientific framework for the geographical discipline. Thus, theoretical geography came into being in which theory is the axis and orientation of the research and explanation of geographical phenomena (cf. Chorley and Hagget 1967; Harvey 1969).

In the field of cartography, Jeremy Crampton (2010) documents how mapping became scientific during the neo-positivistic period. He also reports the influence of Arthur Robinson, and his links with prominent geographers such as Richard Hartshorne, and their work at the Office of Strategic Services (OSS) of the USA during the 1970s. Crampton mentions how by Robinson's contribution, cartography acquired the status of a scientific discipline, a science.

> Perhaps Robinson's best known contribution is his development of the map as a communication system. This focus had the goal of improving the efficiency and functionality of maps as communication devices via empirical experimentation (Crampton 2010: 54).

This statement highlights the neo-positivistic context to which cartography belonged. During the same period, perceptual and cognitive studies of mapping and map use were also initiated (Vanecek 1980). In addition, there were attempts to introduce semiology, modelling theory, and cognition theory to cartography. Thus research and methodologies are framed within the epistemological viewpoint of sciences.

In his article "The Science of Cartography" Árpád Papp-Váry mentions several factors which have accelerated the development of map-making and map production after the Second World War. He calls these factors *the demand for cartography* (cf. Papp-Váry 1989). Papp-Váry describes the evolution of cartography as a science in terms of structural units.[19] He considers how the situation in cartography matches each of the requirements for a new science:

> The subject of cartography [...] can be described as the study of geographic space or of the graphical manifestation of spatial phenomena. The object of cartography is to produce maps which are able *to reflect reality as exactly as possible* (emphasis added, Papp-Váry 1989: 104).

[19] All sciences can be described as consisting of the following structural units: (i) The subject matter of the science; (ii) The manifestation of the subject and its aims; (iii) Characteristic and particular methods of the science and its terminology; (iv) Professionals involved in the science; (v) The institutional organisation, and (vi) Social relationships within the science (Pàpay 1983: 104).

The above statement shows the epistemological perspective characteristic of cartography during the neo-positivistic era. In the object-image-subject relationship, reality (the object) is conceived independently from the subject. The image (map or representation) which the subject grasps from reality must be *accurate* and *transparent,* and *free of value* by the subject (cartographer or mapmaker). Cartography is adapted to the parameters of the new science. Thus the objects of study, the research objectives, methodologies, and results, are framed in the epistemological and philosophical project of the modern period just as in geography and similar sciences.

As Lorenz Hurni and Gerrit Sell interpret John Krygier (1995), the literature about the relationship between art, science, and cartography holds three approaches. As for the first, one sees a polarity, a clear separation between the two domains. The second is a progressive approach which postulates a distinction between the domains, but nevertheless a closer interconnection. It sees science as playing a progressive part, and art as a progressive part in maps. The third approach is also progressive, but offers functional differences for art and science in maps (Hurni and Sell 2009). About the idea of a functional synthesis of art and science in maps, see Krygier (1995). Also, in order to bridge the gap between science and art in cartography, an International Symposium on "Cartography and Art" was organized in Vienna in 2008 (for more detail see Cartwright et al. 2009).

The controversy is rising because cartography has traditionally applied art, science, and technology to map-making to design and realise products. However, the theory and methodology associated with visualising geography has focused on science and technology, and away form art (Cartwright et al. 2009).

Kenneth Field believes that the purposeful combination of art and cartography has put a new agenda in cartography that not only provides a serious contribution to the academic study and appreciation of art and cartography as allied endeavors, but also helps to re-affirm the absolute importance of ensuring that design and aesthetics are central to effective mapping (Field 2009). Along the same lines, highlighting the link between art and technology, Manfred Buchtoithner and Klaus Haberman analized the state-of-the-art of autoestereoscopic relief representation in cartography considering maps, but also true-3D maps, not only as pieces of hard- or softcopy geo-information, but as pieces of art (for more details see Buchroithner and Habermann 2010).

Additionally, in his a little epic paper "Im Anfang war die Karte" ("In the Beginning was the Map"; drawing from Saint John 1:1 "In the Beginning was the word."), the Italian geographer Franco Farinelli goes back to Anaximander of Miletus (610–546 BC) and Immanuel Kant who were, before turning to philosophy, geographers. Kant as well as the German mathematician, logician and philosopher [Friedrich Ludwig] Gottlob Frege (1848–1925) are drawing on Anaximander when they are questioning the sense of the reduction of our geographic discoveries and insights to a description in the form of a cartographic depiction (Farinelli 2013). In fact, only during the second half of the 20th century "the map eventually succeeds to convert into a photography and to assimilate the model by means of airphotos and satellite images nearly to a complete amalgamation."[20] (Farinelli 2013). CIL maps ("combined image-line maps) are increasingly replacing classical topographic maps (Buchroithner and Kostka 1994, 1996, 1997). The advent of classical airphotos provided the basis for elevation contours and, initially, for triangulation (at the turn from the 19th to the 20th century): a scheme of "the inner relations of the depiction of an object by means of signs/symbols[21] the most differentiated scheme" according to Farinelli (2013) we are currently disposing of in cartography.

It is remarkable that nobody less but the Austrian philosopher Edmund [Gustav Albrecht] Husserl (1859–1938) postulates that, as far as drawings of geo-relations are concerned, one *has* to draw these figures oneself (Husserl 1996: 53). This, again, implies that every geometric construction represents a causal system, a hierarchy between causes and effects (Farinelli 2013), i.e. amongst others between terrain and map. To "draw" yourself is the only means not to incur adverse effects but, on the contrary, to control them. This might also explain why the texts of critical geographers like Alexander von Humboldt or Carl von Ritter do not contain a single map. According to Farinelli (2013: 275) cartography decides implicitly and tacitly about (not only the place but also) the *nature* ("*Wesen*") of things, appealing to the absolute power of maps which "allow neither critique nor corrections".[22]

1.5 Postmodernism and Poststructuralism

Postmodernism is a term of cultural character which arose in the last quarter of the twentieth century. This term designates the radical cultural change of features identifying the end of *modernity* (Ortega Valcárcel 2000).[23] It questions the scientific, epistemological, cultural, and ideological assumptions which support the development of the Western culture since the age of Enlightenment. This questioning is the most striking feature of postmodernism.

For postmodernists, knowledge and truth are always relative to a particular culture or to a historical period. According to William Gorton, this applies not only to moral and aesthetic judgments, but also to claims for truth made by the natural and social sciences (Gorton 2010).

---

[20] "Erst während der zweiten Hälfte des letzten Jahrhunderts gelingt es der Karte, die sich in eine Photographie verwandelt, sich vermittels Luft- und Satellitenbildaufnahmen ihrem Modell fast bis zur Verschmelzung anzugleichen." (Farinelli 2013: 271).

[21] "Die Triangulation … [brachte] … noch ein Schema von den inneren Verhältnissen der Darstellung eines Gegenstandes vermittels eines Zeichens hervor – das differenzierteste Schema, über das wir verfügen." (Farinelli 2013: 271).

[22] "…eine Kartographie, … in welcher sie sich eben genau auf die absolute Macht der karte berief, die weder Kritik noch Korrekturen zulässt."(Farinelli 2013: 275).

[23] Here Ortega Valcárcel is drawing on the work of Friedman (1989).

> Thus science does not offer a method for arriving at universal, objective truths that transcend time and place. Rather, it represents one way of knowing that reflects certain values, beliefs and interests of modern, Western society. Moreover, for postmodernists there is no fixed, universal human nature. Instead, human nature (our beliefs, values, desires, interests, and even our emotions) is itself a product of a particular history or social configuration – or, as postmodernists sometimes say, human nature is *socially constructed* (original emphasis, Gorton 2010, digital text: no pagination).

The postmodernists reject the uncovering of patterns, structures, or laws that purportedly transcend history and culture as deeply misguided attempts by social scientists. These attempts were made in the nineteenth century, especially during the validation of positivism as applied in social sciences (for more details see Gomez Mendoza et al. 1988). For postmodernists, the understanding of a particular society must be local and contextual.

According to José Ortega Valcárcel a theoretical criticism arises during this time which analyses the incongruities and contradictions of the philosophies that govern the cultural, social, scientific, philosophical, and epistemological patterns of modern society. It is a criticism directed towards the foundations of modernity. This is called *poststructuralism.*

> The corroboration of a new time and a new culture is formulated. The new culture defines itself as *postmodernism.* The new time defines itself as *postmodernity*. Theoretical criticism, or poststructuralism, and a new culture or postmodernism make up postmodernity[24] (original emphases, translated from Ortega Valcárcel 2000: 242).

The postmodern culture is sustained on the criticism of modernity. This criticism begins with the Frankfurt School (founded in Germany), which arose towards the end of the first third of the twentieth century, and acquired relevance after of the Second World War. The main thinkers of this school are: Theodor W. Adorno (1903–1969), Herbert Marcuse (1898–1979), Walter Benjamin (1892–1940), Erich Fromm (1900–1980), Max Horkheimer (1845–1973), and Jürgen Habermas (born 1929) (Bretow 2002; Gorton 2010; Ortega Valcárcel 2000). This intellectual movement criticises modern capitalism and its theoretical and epistemological support. The original critical theorists argued that a social scientist should not—and cannot—be a neutral observer of the social world. In this way, 'in place of orthodox Marxism they aimed to produce a new theory that could at once explain the failure of socialism in the Western liberal democracies and also provide a critique of what they saw as oppressive features of developed capitalist societies' (Gorton 2010).

Science and technology do not escape this critical movement:

> The interpretation of the capitalism from the perspective of the *domain* constitutes a fundamental feature of the critical conception of this *school*. In accordance to this, science

---

[24] 'Se formula como afirmación de un tiempo nuevo y una cultura nueva. La nueva cultura se define como *posmodernismo*. El tiempo nuevo corresponde a la *posmodernidad*. Crítica teórica, o postestructuralismo y nueva cultura o posmodernismo, configuran la posmodernidad'.

> and technology constitute the axis and the support of this domain[25] (original emphases, translated from Ortega Valcárcel 2000: 244).

From this intellectual movement, other authors who were related to the European political left also became prominent during the second half of the twentieth century. A French intellectual group in the field of culture and social sciences included authors such as Michel Foucault, Jacques Derrida, Gilles Deleuze and Félix Guattari, Jean Baudrillard, and Jean-Francois Lyotard. They all doubted the assumptions held since the Enlightenment.

Deleuze and Guatteri (both referring to the viewpoints of philosophy and psychoanalysis) investigated the relationship between capitalism and mental disorder and between capitalism and desire. In other words, they claimed a relation between the social system and the individual impulses. Foucault formulated equivalent conclusions regarding the relationship between power and knowledge and stated that 'there is no truth outside of power' (Ortega Valcárcel 2000: 245). Further, he linked the truth, i.e., objectivity, with the social horizon. Derrida tackled the relationship between language and thought. The fundamental idea is that language shapes reality; even more, language is reality. Finally, Lyotard set out to explain *narrative knowledge*[26] and he poses that science is a 'subset of knowledge' (Woodward 2005, digital text: no pagination).

Other philosophers in the context of the history of science and epistemology, like Thomas Kuhn, take the truth of scientific knowledge as *relative* because it is socially conditional. Therefore, this *scientific truth* does not exceed the status of discourse (Kuhn 1970; Hall 2006):

> Poststructuralism as criticism of the rationality of the Enlightenment is considered. It supports an intellectual trend with authors like J. Baudrillard and J. F. Lyotard which marks antirationalism. It has to be distinguished from the denunciation of the scientific discourse. Structural theories and conceptions of universal character are rejected. The assumptions on which the modern world has been constructed are denounced. This means, the rational subject, reason, and scientific knowledge are identified with the truth[27] (translated by the authors from Ortega Valcárcel 2000: 247).

Table 1.2 presents a schema comparing neopositivist epistemology and post-positivist epistemology of sciences (Romaniuk and Paillalef 2010). The former

[25] 'La interpretación del capitalismo desde la perspectiva del *dominio* constituye un rasgo fundamental de la concepción crítica de esta *escuela*. De acuerdo con ella, la ciencia y la técnica constituyen el eje y el soporte de ese dominio'.

[26] For Lyotard, narrative knowledge is the kind of knowledge prevalent in primitive or traditional societies, and is based on storytelling, sometimes in the form of ritual, music, and dance (from Woodward 2005).

[27] 'El postestructuralismo se perfila como una crítica a la racionalidad de la Ilustración. Alimenta una corriente intelectual en la que destacan autores como J. Baudrillard y J.F. Lyotard, de acentuado antirracionalismo. Se distinguen por la denuncia del discurso científico. Rechazan las teorías estructurales, las concepciones de carácter universal. Denuncian los presupuestos sobre los que se ha construido el mundo moderno, es decir, el sujeto racional, la razón y el conocimiento científico, identificado con la verdad'.

Table 1.1 Different *isms* dealing with the relationship between subject–object in the context of the Theory of Knowledge Adapted from Hessen (1976)

Human knowledge	Subject–object relationship	Epistemological and philosophical perspectives ('isms')
Possibilities of human knowledge	Can the subject really *capture* the object?	• Dogmatism • Subjectivism • Pragmatism • Criticism
Origin of human knowledge	Is *reason* or *experience* the source and base of human knowledge?	• Rationalism • Empiricism • Intellectualism • Apriorism
Essence of human knowledge	Is it the *object* that determines the subject or is it the *subject* that determines the object?	• Objectivism • Subjectivism • Realism • Idealism • Phenomenalism

corresponds to the *modern* period and the latter one belongs to the philosophy of the *postmodern* period. The left column of the table shows the aims, methods, and contexts established by the Vienna Circle which were valid during the predominance of logical positivism. The right column contrasts the new conceptions regarding the same aspects mentioned above, in the context of postpositivism, especially that of the second half of the twentieth century and even at the time of this publication (Table 1.1).

On the other hand, the postmodern approach is not new. It is a fresh outbreak of the great antirationalistic movement of the end of the nineteenth century and the first third of the twentieth century that was known as *historicsm*[28] (Valcárcel 2000). The roots of postmodernity correspond to the *philosophies of the subject*, because in historicism the importance is given to the subject (or the individual; cf. Ortega Valcárcel 2000). Therefore, when reading the distinctive features of the foundations of postmodernism or poststructuralism in the right-hand column of Table 1.2, characteristics of historicism can be seen. For instance, the recognition and study of the links between science and its historical context is a feature of historicism.

1.5.1 Postmodernism and Cartography

Cartography and mapping, as science or as a discipline in the scientific context, have not escaped postmodern criticism. This is known as the contemporary *critique of cartography*. Nikolas Huffman has, from the standpoint of map design in

[28] *Historicism* is a mode of thinking that assigns a central and basic significance to a specific context, such as historical period, geographical place and local culture Kahan (1997).

Table 1.2 Comparison between the neopositivist and postpositivist approaches in epistemology and philosophy of sciences. Translated from Romaniuk and Paillalef (2010)

Neopositivist Epistemology (Vienna circle)	New philosophy of science (Postpositivist)
To unify all sciences with one unique method	To pay attention to the diversity and peculiarities of every discipline
To formalise the language of science (without considering the use contexts, and sometimes not even the meaning)	The meaning depends on the context; interest in the practices of scientists
Human rationality is equivalent to scientific reason	Extends the concept of rationality
Ideal of an ethical neutrality of science	To recognise and study links between science and society, between science and its historical context, between science and psychological processes, etc.
Attention to the methods and language of science and the logical aspects of theories	Focus on the practices of scientists, how the scientific communities behave, how decisions are made, etc.
Distinction between discovery context, justification context, and application context	Distinction of contexts is artificial

postmodernism, analysed maps and mapping within the postmodernistic framework. He addresses the postmodern critique by outlining four different definitions of postmodernism and its relation to contemporary cartographic critiques. These categories are: mapping and postmodern style; cartography and postmodern social theory; capitalism and the economy of mapping; and cartography and poststructuralism (for more details see Huffman 1996).

Huffman mentions some postmodernists in cartography such as Denis Wood, J.B. Harley, Richard Helgerson, Barbara Belyea, and Robert Rundstrom. These authors question the apolitical and scientific status of cartography.

> These critiques have disputed the way that language and the production of meaning have been theorized in cartography research on maps and mapping, and introduced *new ways of understanding* how we interact and communicate with and through maps. [...] They have also pointed towards a broader sociology of mapping in which maps and mapping can be understood as *artefacts within our social and material culture*, and have demanded that greater attention be paid to issues of representation, politics, and social action (emphases added, Huffman 1996: 35–36).

The above statement points out new tendencies in contemporary cartography which fall within the postmodern context. These new ways of understanding maps are referred to as being an alternative to the scientific-empirical approach in cartography during modernity. Thus the map is an artefact (material or ideal) within the social and cultural context in which it is created and used, and is no longer a device with an objective, neutral, and value-free character. In other words, all these new visions in cartography and mapping are framed within the new postpositivistic epistemology of sciences.

In his analysis of cartography from a postmodern point of view, Huffman (1996) mentions that cartography has been influenced by postmodern social theory, especially by Foucault's social critique (1969), which was also adopted by Harley (1988a, b, 1989, 2001) and applied to cartography. 'Concepts like *power-knowledge*, *episteme*, and *discourse*, were applied to the role of maps in creating and sustaining social and political power within a given society' (Huffman 1996: 39). The author also refers to Barbara Belyea's (1992) reworking of Foucault and cartography, and Latour's recent work in the sociology of sciences (Latour 1990). Denis Wood claims that 'maps are *weapons of power* that create the territory desired by those empowered to make and enforce maps' (Wood 1992).

When Huffman analyses cartography and poststructuralism, he highlights a "crisis of representation" created by the postructuralist philosophy of language.[29] His 'work has focused on the language of maps as complex social and cultural objects and on how maps construct and are constructed by society and language' (Huffman 1996). He mentions Derrida's philosophical critique and the adoption of Derrida's approach by J.B. Harley who analyses *maps as texts* 'that seek[s] to reveal the underlying political interests and prejudices embodied in maps' (Harley 1989).

The section on "Critical Cartography in the Context of Post-Modernism" provides a deeper analysis of Harley's legacy and his drawings from Foucault' power-knowledge relationship as well as Derrida's hermeneutics applied to cartography. In this way, the poststructuralist and postmodern approaches in the field of cartography, conceived as a *social activity* and maps as *devices of power and persuasion*, are seen with a different perspective in comparison to that of the modern period of cartography.

1.5.2 Mental Map Generation Through Map Reading: Against Ubiquitous Geoinformation

Since the advent and the enthusiastic acceptance of ubiquitous cartography, there exist also negative voices cautioning against a general loss of spatial conception. As stated by the German psychotherapist and professional mountain guide Martin Schwiersch, with the emergence of the "panacea" of satellite-based navigation our "spatial knowledge" is definitely about to disappear (Schwiersch 2011). Manfred Buchroithner wrote about the "loss of our cognitive map" (Buchroithner 2012). In modification and continuation of the famous French naturalist's and

[29] Huffman summarises the crisis of representation in two major tenets. "First, all human language and meaning is based on the interpretation of signs that are ultimately indeterminate and always open to further analysis. Second, [...] all human knowledge [...] is mediated through language structures that shape our perspectives on the world around us, as well as constituting such personal concepts as self and subjectivity" (Huffman 1996: 41). This poststructuralist position has important repercussions for the map's conception.

mathematician's René Descartes's sentence from 1644 "cogito ergo sum" ("I think, therefore I am", initially 1637 in French "Je pense, donc je suis.") and drawing on Immanuel Kant but also on the critical statements of Jaakko Hintikka and Rudolf Carnap (Carnap 1931) he claimed that if, in difficult terrain, one does not *know where I am* ("*cogito ubi sum*"), one's being can easily be jeopardized or even find an end. A 'virtual landscape in our brain' should again replace the loading of GPS coordinates onto our smart-phones or other mobile devices (Buchroithner 2012: 31–32).

The blessings of *location-based services (LBS)* are indisputable (cf. Kainz et al. 2013, Buchroithner and Gartner 2013). Their current and future possibilities and their role in resp. for modern cartography are in a summarising manner described by Gartner 2013. On the other hand, today (2013) every carrier of a mobile phone can, at least roughly, be located; a blessing in the case of an accident. But, do we really want to be localizable at every second, everywhere? Professional drivers are increasingly using *jammers*, i.e. transmitters interfering with the GPS signals, in order not to be caught by the fleet managers when making small deviations or little nap-stops. Not only the military but also various land and sea authorities are asking what would happen if whole regions are GPS-wise incapacitated. Alone in summer 2011 more than one-thousand jammer incidents a day have been registered by the U.S. army anti-jammer network *J-lock*. 'Many of these interfering signals simply stem from defect electric devices.' (Schmundt 2011a).

With respect to navigation, David Last, one of the most influential European GPS experts, said 'It would be unjustifiable to further only build on satellite signals.' (Schmundt 2011b: 129). 'The deficient set-up of a mental spatial model, which is caused by the use of navigation systems, is for the time being the price to pay for an easy route search between two places.'[30] wrote Dickmann and Kestermann (2011), based on statistically sound tests using both maps and navigation systems in an urban area.

When stating that 'Today *cogito ubi sum* has [...] again to gain the importance of classical "map reading" and the acquisition of "space knowledge" in the true sense of the word.'[31] Manfred Buchroithner might be considered spokesman of a technology-obsessed but critical number of cartographers (cf. Buchroithner 2012).

---

[30] Der defizitäre Aufbau eines mentalen Raummodells, der infolge der Nutzung von Navigationssystemen entsteht, ist vorerst der Preis für eine vergleichsweise mühelose Routensyuche zwischen zwei Orten. (Dickmann & Kestermann 2011: 188).

[31] *Cogito ubi sum* muss heute [...] wieder jene Bedeutung des klassischen „Karten-Lesens" und des Gewinnens von „Raum-Wissen" im wahrsten Sinne des Wortes bekommen. (Buchroithner 2012: 37).

Chapter 2
A Philosophical Framework Applied to Cartography

2.1 Introduction

In this chapter we want to refer to one of the most prominent philosophers: Immanuel Kant. In a cartographic context his contributions to the field of geography and the concepts of *space and time* are very important. For Kant, space as well as time are only "a priori" concepts to understand phenomena, similar to intuition. Kant postulates that space and time are not real, but only a sensible projection of the sense of symmetry of our own corporality and of our own sense of change with the objective to put all phenomena into an order.

Further, two great philosophers of logical positivism are analysed: Ludwig Wittgenstein and Karl Popper. We are going to ask how cartography grasps epistemological aspects of knowledge construction. Neither Wittgenstein nor Popper wrote explicitly about cartography and mapping, but their legacy has an important impact on our understanding of maps. The evolution of contemporary cartography from an epistemological point of view can be considered to be equivalent to Wittgenstein's trajectory (his First and Second Philosophy). On the other hand, cartographic products and the different stages of map creation can be linked to Popper's Three World Theory.

2.2 Immanuel Kant

2.2.1 No Epistemology Without him: Kant's Role for Spatial Sciences

Having lived at a time of major successes and advances in natural sciences by individuals like Robert Boyle, Alexander von Humboldt, Carl Friedrich Gauss, and Isaac Newton, Immanuel Kant (22 April 1724 – 12 February 1804) as simply the most famous German philosopher and one of the most influential of all times, cannot be neglected when writing about spatial facts and their graphical depiction. Kant proposed that human knowledge could be organised in three ways: by

P. I. Azócar Fernández and M. F. Buchroithner, *Paradigms in Cartography*,

DOI: 10.1007/978-3-642-38893-4_2,

classifying facts according to the type of the objects studied, by examining the temporal dimension and looking at things in terms of their history, and by understanding facts relative to (their) spatial relationships. The latter fact represents nothing but the field of knowledge commonly known as geography (Amodeo 2005).

Further, Kant divided geography into the sub-disciplines of physical, moral, commercial, and theological geography. He described geography as a discipline that synthesises the findings of other sciences through the concept of space. He taught geography for 30 years, explaining each sub-disciplines by clarifying the position of geography among the many fields of learning (cf. Buchroithner and Azócar 2011). The idea that geography deals with the differentiation of places was fundamental to Kant's understanding of the world. He also saw a clear distinction between history and geography (Amodeo 2005). "Nichts bildet und kultiviert den gesunden Verstand mehr als Geographie" ("Nothing educates and cultivates the common sense more than geography.") This statement by Immanuel Kant made during his lectures in Physical Geography is bequeathed through the handwritten notes of one of his students and clearly shows the importance he assigned to the science of geography (Stark and Brandt 2009).

When Kant gave classes on physical geography, he focused on the interconnection of phenomena as a way of causally explaining them in space and time. In Kant's philosophical framework, his lectures represented the result of thought about the differentiation and movement of the mentioned phenomena.

Kant is also important for cartography because he involves perception in mapping in a twofold sense: for the mapping and surveying in the field (respectively, nowadays, by remote sensing) and for the perception of ready-made maps. Kant defines his theory of perception in his influential 1781 work *The Critique of Pure Reason* (*Kritik der reinen Vernunft*), which has often been cited as the most significant volume of metaphysics and epistemology in modern philosophy. He maintains that our understanding of the external world had its foundations not merely in experience, but in both experience and a priori concept, thus offering a *non-empiricist critique of rationalist philosophy*, which is what he and others referred to as his "Copernican Revolution"(cf., e.g., "Kant, Immanuel" in The Columbia Encyclopedia, Sixth Edition).

2.2.2 *Kant's Philosophy: An Epistemological Frame for Geography and Cartography*

José Ortega Valcárcel analysed Immanuel Kant and Geography within an epistemological framework. For him, Kant as geographer does not initiate modern geography, but rather culminates the representation of the earlier medieval world. Nevertheless, Kant's postulates on human knowledge influence the conception of space and geography of modern geographers. The theoretical-methodological considerations that Kant develops work like an introduction with regard to the human knowledge, its forms, origins, and classification, and have had notable aftereffect (Ortega Valcárcel 2000).

For Kant, the process of arranging our experiences as knowledge, that is, the rational process, takes place in accordance with concepts or according to time and space. Kant calls the classification of knowledge according to concepts a *logical classification*. He names classification in accordance with time and space *physical classificacation*. The first one represents a natural system, (like e.g. the one by Carolus Linnaeus); the latter one, a physical description of nature.

Ortega Valcárcel concludes that Kant framed geographical knowledge in this way. He called it the "physical description of the Earth", in other words, "physical geography" (Ortega Valcárcel 2000: 111). For Kant, the physical description is the foundation of knowledge of the world. The world is the substratum and foundation on which our knowledge must arise.

The above statements lead to the first manifestation of *formal cartography*, i.e. the cartographic representation of the physical world through topographic or reference maps. These types of cartographic products were created firstly, to depict the physical elements and processes of the world. Here one can refer to the metaphor of *the map as a reflexion of nature*. Only subsequently were other elements belonging to the social, economical, and political world (i.e. thematic map) been integrated. But the traditional substratum in cartography is physical. All the other components of reality are constructed on cartography's physical layer.

As mentioned in various books about Immanuel Kant, he himself stated in 1800 that he had been giving lectures in Physical Geography "for some thirty years" ("einige dreißig Jahre hindurch"). Needless to say, Kant is one of the most influential scholars of the science of geography. In his seminal book "History of the Geographic Science" ("Geschichte der geographischen Wissenschaft von den ersten Anfängen bis zum Ende des 18. Jahrhunderts") Josef Schmithüsen (1970) devotes as many as eight densely printed text pages to the role of Kant in the science of geography. However, no single statement can there be found about cartographic depictions of geo-phenomena. Also an in-depth search in the literature about Kant will not yield any findings about his importance for cartography (cf. Hartshorne 1958; Nesher 1997; Amodeo 2005).

2.3 Ludwig Wittgenstein

2.3.1 Wittgenstein's Early Work "Tractatus logico-philosophicus"

Ludwig Wittgenstein (26 April 1889 – 29 April 1951) is considered to be one of the main philosophers of all times who influenced development of philosophy since the twentieth century (cf. i.a. Schulte 2005). Although he was not a member of the Vienna Circle, Wittgenstein's philosophy belongs to the logical positivistic approach. He published his contribution to the history of thought in two important and controversial books: "*Tractatus logico-philosophicus*" (English version 1922) and "*Philosophical Investigations*" (1953; Richter 2004a). Much has been written in philosophy about these two books of Wittgenstein (Monk 2005).

Wittgenstein never directly addressed philosophy issues regarding cartography and mapping. Nevertheless, the philosophical and epistemological aspects of his reflexions in relation to cartography will be discussed in this chapter. Here, any further considerations are only suggestions that are subject to deeper analysis.

According to Wittgenstein, the relationship between language and the world can be presented through a model *by proxy*.[1] Thus, a relationship by proxy between a model of reality and the reality itself is established. These claims are of critical importance for understanding the world through language. The most important of Wittgenstein's contributions to the world knowledge is: *language is a model of reality*, and reality is comprehended by us through language. Similarly, if these insights are critical to Wittgenstein's thought, then cartography, *conceived as a model of reality*, will also have important epistemological considerations. In this sense there are many visual models to depict reality in cartography. Consequently, these models establish a relation by proxy between their components and the external elements which are represented in a map form (relation of object to image).

Therefore, a general reading of Wittgenstein's first works, more precisely in his *Tractatus*, leads to understanding a relation between its content and the nature and objectives of cartography. Several of the propositions defined by Wittgenstein can be applied to cartography and mapping, because one of the traditional objectives in cartography is to represent reality (i.e. to depict) and primarily to depict the physical objects of the world. This representation is made through certain material and digital devices. Similarly, there is also the parallel creation of an image of the external world, but this image is inside our internal worlds of the mind.

The map is then a representation of reality, and the map must be created. There is a one-to-one (biunique) correspondence between what is represented and the reality: the symbol on the map represents in this case the objective element that belongs to reality. In general terms, Wittgenstein's early philosophy assumes a coincidence with the first stage in the development of modern cartography: In both the positivistic and neo-positivistic context, the map is considered to be an objective, accurate, and genuine means to depict the real physical world. The most important contributions of the *Tractatus logico-philosophicus* are the *picture theory of meaning* and the *doctrine of logical atomism.*[2] These theories are intimately connected. The picture theory states that language draws a picture of reality (Richter 2004b). The pictorial nature of thought and language is analysed by Pasquale Frascolla as a way to understand Wittgenstein's approach. Essentially this picture theory of meaning states that "our language and our thought have

---

[1] A *proxy* is generally understood as a person who represents another person. In other words a person authorised to act for another (Thesaurus Dictionary year?). This analogy is applied here to the relationship between language and world.

[2] Bertrand Russell conceives *logical atomism* as the view that reality consists of a great many ultimate constituents or "atoms". "Logical" atomism is an attempt to arrive through reason at what must be the ultimate constituents and forms constituting reality (Carey 2008).

sense and reference, because there are paintings, figures, or representations of the things of the world" (Frascolla 2007).

From an epistemological perspective, Wittgenstein's main contribution is not only about language, but about a theory of the world, namely the *theory of knowledge of the world*. In his atomistic view, Wittgenstein claims that the world is composed of facts, states of affairs, and objects, each one having a correspondence in language: propositions, elements of propositions, and names respectively. Names refer to the objects of reality, and hence *the meaning of the object is in its reference* (Clack 1999).

Some important and representative propositions from the *Tractatus* (cf. Richter 2004a, digital text: no pagination) are[3]:

1. The world is all that is the case.
4.01. A proposition is a picture of reality.
4.0312. ... Propositions *show* the logical form of reality. They display it.
4.5. ... The general form of a proposition is: This is how things stand.
5.4711. To give the essence of a proposition means to give the essence of all description, and thus the essence of the world.
5.6 *The limits of my language* mean the limits of my world.

If language is a perfect analogy of the world, the *cartographic language*—in the context of map symbols—is an important epistemological contribution to the theory of knowledge. According to Yu Liansheng, map symbols belong within a category of scientific symbols, and at the same time possess the features of visual images. He describes the philosophical aspects of map symbols and the exploration of their information function (Liansheng 1997). Consequently, the essence of a map symbol, its characteristics, its poly-functions, and its informational function are all perfectly related to the one-to-one relationship established by Wittgenstein between language and world (here, as already mentioned above, the world is conceived as reality).

On the other hand, within the context of information exchange among people, Pavel Neytchev points out similarities between the units of natural and of map languages in the realms concerning the syntactic components of cartographic sentences. He claims that the map language is a double-articulated code, and later on defines the syntactic patterns of cartographic sentences (Neytchev 2001). Neytchev's contribution can be perfectly compared to Wittgenstein's claim, presenting the function of the cartographic language in the context of grammar and syntax. There is a connection among "map language" (Schlichtmann 1985, 2009; Ramírez 2004; Neytchev 2001), natural language, and knowledge of the world or reality.

Both authors, Liansheng and Neytchev, are important to understand the cartographic language about the knowledge of the world. Liansheng establishes the philosophical levels of map symbols and the exploration of their information

[3] The numbers at the beginning of the listings correspond to the respective chapter and subchapter classification in the *Tractatus logico-philosophicus* (cf. Wittgenstein 1921, 1922, 2005).

Table 2.1 Knowledge of the world and language according to Ludwig Wittgenstein

World	Language	Units of map language	
		In text reading	In map language use
Objects	Names	Separate cartographic sign (cartographic word)	Component of cartographic utterance
States of affairs	Elementary propositions	Compound cartographic sign (cartographic sentences)	Cartographic utterance
Facts	Propositions	Map (cartography textual work)	Cartographic *communiqué*

Adapted from Clack (1999) and Neytchev (2001)

function by analysing the essence, characteristics, function, and laws of operation from ideological, cultural, and philosophical considerations. Neytchev compares the units of natural language (in linguistic systems, in texts, and acts of speech) to units of map language (in cartographic systems, in text, and in map language). An important aspect of Neytchev's work is the use of units of language: basic sign, combined sign, and assembled sign. They correspond to the components used by Wittgenstein in his description of the elements that belong to language: propositions, elementary propositions, and names (see Table 2.1).

2.3.2 *Wittgenstein's Late Work* "Philosophical Investigations"

In the second phase of Ludwig Wittgenstein's work (also called Wittgenstein's later philosophy), he carried out a critical analysis of his first major monograph, the "*Tractatus*". This critique is compiled in the book "*Philosophical Investigations*" published in an English translation of the original German "*Philosophische Untersuchungen*" by Elisabeth Anscombe in 1953, two years after of the philosopher's death. Although this later phase of his work was more remote from science, it was a source of inspiration for many philosophers and scientists because of its scope and considerations (cf. Richter 2004b).

This part of Wittgenstein's philosophy is considered by his followers to be postmodern.[4] Wittgenstein rejects the supremacy of *declarative language* and essentialistic vision of the language. Using language, more things can be done than

[4] *Postmodernism* is a term that designates a wide number of artistic, cultural, literary, and philosophical movements of the twentieth century which are critical and in conflict with the modernist period.

only describe the physical world. In this way, Wittgenstein considers numerous problems and puzzles in different fields such as semantics, logic, philosophy of mathematic, philosophy of psychology, philosophy of action, and philosophy of the mind. His main contribution is the idea of *language-games*, namely that language functions in the context of known-rules; all prepositions that are out or beyond these rules are inconceivable and meaningless. The meaning and reality of the world belongs to several contexts; therefore, descriptive language is only one aspect. In this context, the aim of philosophy for Wittgenstein is to clarify these *rules* or *language-games* for the understanding of the world. Consequently, all concepts which fall outside of these rules are considered to be contradictions, antinomies, meaningless, or senseless (Richter 2004a). This part of Wittgenstein's work is more flexible than the rigid approach of his *Tractatus logico-philosphicus* where, in the ideal of language, a meaning must correspond to every word and vice versa. Now, the meaning of the words depends on the context.

Since the 1990s, important contributions in cartography have been coming from sources outside the discipline, or are frequently not unique in the academic context. Therefore, these contributions are framed within the deconstructionist or post-structuralist approaches. In this sense, there arises strong criticism of how mapping has been practiced and developed until the end of the 2000s. The main exponent of this development is J.B. Harley who considered cartography to be far from an objective and accurate discipline, as is regarded by the positivistic approach. On the contrary, cartography is full of subjectivities, and maps, in particular, are full of intents and inaccuracies. The meaning of maps is, however, valid according to the social context in which they are used or incorporated (Harley 1989).

This new movement in the development of cartography is called *critical cartography*. For Crampton and Krygier (2006), this trend challenges academic cartography by linking geographic knowledge with power. During the post-war period, cartography underwent a significant solidification as a science, while at the same time other mapping practices were occurring. The authors focus their analysis on the critical theoretical and *critical mapping practices* in historical perspective.

At this point, it is also important to consider Tomasz Zarycki's pragmatic approach to map analysis. He points out obvious differences between *map semantics* and *map pragmatics* in the context of traditional divisions of semiotics (semantics, syntactics, and pragmatics).[5] He claims that while the semantic analysis of maps will concentrate on the extent to which the criteria of the objective map-making—or *the rules of objective representation*—are fulfilled by particular maps; *pragmatic analysis* should concentrate on establishing the nature

[5] *Semantics* deals with the meaning of the symbols (relationships between the sign and vehicle or referent); *syntactic/syntax* deals with the formal proprieties of signs and symbols (relationships sign-vehicle/sign-vehicle); and *pragmatics* deals with all the psychological, biological, and sociological phenomena that surround the functioning of cartographic signs (relationships between the sign and vehicle or interpreter). *Referent* is the object. *Interpreter* implies the concept (thought or reference). Extracted from Freitag (2001), Kavouras and Kokla (2008), and Gartner (2009).

of actual social contexts and other criteria of acceptability of maps (cf. Zarycki 2001a).

In Zarycki's map semantics approach, he writes that the 'map is conceived as a tool for the description of reality'; on the other hand, in the map pragmatic approach, the 'map is conceived as a form of creation/negotiation of reality'; the 'map is considered to be a tool of symbolic domination'. The 'rules of cartographic communication are objectively given and must be respected' in map semantics; whereas in map pragmatics the 'rules which govern the cartographic communication are unstable and must be established, analyzed and related to some social context of their existence'. Finally, in the map semantics approach, 'maps are created by a cartographer on the basis of his/her knowledge about reality'. In the map pragmatic approach 'maps are designed not only by those who make them but also by the interests of those whom they serve. The map appears to be under the direct and indirect influence of the potential or actual users' (Zarycki 2001a: 69).

When Zarycki describes the characteristics of the semantic map and of the pragmatic map, there is a connection between Wittgenstein's earlier approach and the later approach described by Wittgenstein. This means that the features of map semantics belong to the strict criteria of objectivity when Wittgenstein sets up a one-to-one relation between language and reality. Nevertheless, the conceptions of map pragmatics are different. These elements coincide with the evolution of thought of the later writings of Wittgenstein, when he claims that the descriptive language is only one part of the different kinds of languages. The maps are pragmatically analysed in the context of post-structuralism and social theory,[6] which claim other alternatives for seeing and understanding the world.

In summary, Wittgenstein's thought evolution (*first*- and *second Wittgenstein* or his *early and later philosophy*) is manifested in a semiotic approach to cartography when map semantics and map pragmatics are confronted (see Table 2.2). Hence, it can be claimed that the philosophy of Wittgenstein's *Tractatus*, when applied to cartography, belongs to the scientific and objective cartographic approach. On the other hand, Wittgenstein's thought in *Philosophical Investigations* corresponds to critical cartography,[7] namely to an alternative cartographic approach. The first cartographic approach belongs to the traditional positivism or neo-positivism of sciences, and the second one goes beyond academia, namely post-structuralism or deconstructionism.[8]

---

[6] *Social theory* refers to the use of abstract and often complex theoretical frameworks to describe, explain, and analyse the social world (New World Encyclopaedia).

[7] *Critical cartography* aspects will be analysed in the section of this book called "Critical Cartography in the Context of Post-Modernism".

[8] These perspectives belong to the postmodern period. *Post-structuralism* and *Deconstructionism* will be analysed in the following chapters of this section and in the section "Critical Cartography in the Context of Post-Modernism".

Table 2.2 Parallelism between Wittgenstein's philosophy and the evolution of cartography during the modern and postmodern periods (After Azócar 2012)

Ludwig Wittgenstein's philosophy	Cartographic perspective
"Early Wittgenstein": *"Tractatus logico-philosophicus"* (1921) Key aspect: *Language and World*	Modern cartography Scientific cartography Positivism and Neo-positivism approach of mapping (e.g. semantic maps)
"Late Wittgenstein": "Philosophical Investigations" (1953) Key aspect: *Language-games*	Postmodern cartography Critical cartography Deconstructionism and Post-structuralism approach of mapping (e.g. pragmatic maps)

2.4 Karl Popper

Karl Raimund Popper (28 July 1902–1917 September 1994) was intellectually and ideologically close but actually never invited to participate in the *Vienna Circle*, the Viennese school of logical positivism, (Keuth 2004) and wrote basic books on the philosophy of science (cf. Moritz 1995). One of Poppers' major contributions to the theory of knowledge is about the various worlds of knowledge. Already in 1934 he published his book about 'Logik der Forschung',[9] twenty-five years later published as 'The Logic of Scientific Discovery' (Popper 1934/35, 1959). These two seminal books actually also served as a basis for Thomas Kuhn's reflections. For Popper there exist basically two kinds of knowledge: *subjective knowledge* and *objective knowledge*. Popper, together with John Eccles, introduced the Three Worlds Model in their book *The Self and its Brain* first published in 1977 (Popper and Eccles 1993).

Popper's theory of the three worlds establishes a distinction among the *world in itself*, the *subjective world* and the *objective world*. The world itself remains, in Kantian language, a *noumenon,*[10] that is unknowable to the human. From this world we can only study the phenomena. The Second World is that of the individual conceptions of thought, a completely subjective world, exclusively dependent on the individual's point of view. The Third World is an objective world because of its inter-subject validation of conceptions which initially had an individual character (Mejia Soto 2004).

[9] Imprint '1935', actually already published in 1934.

[10] A *noumenon* may, according to Wikipeadia [include reference], be described as a posited object or event as it appears in itself independent of perception by the senses. Noumenon is the thing in itself, reality per se (it remains unknowable). According to the Theory of Knowledge of Immanuel Kant, presented in his *Criticism of the Pure Reason*, the intellect does not know the things as they are in themselves (*noumena*) but as they construct themselves (*phenomena*).

In his description of the three worlds Popper formulates:

> First, there is the physical world – the universe of physical entities – ... this I will call "World 1". Second, there is the world of mental states, including states of consciousness and psychological dispositions and unconscious states; this I will call "World 2". But there is also a *third* such world, the world of the contents of thought, and, indeed, of the products of the human mind; this I will call "World 3" (original emphasis, Popper and Eccles 1993: 38)[11]

In this way, World 3 'is inhabited by the set of products of all our cultural activities and comprises all human works from the point of view of their logical and objective content' (Gattei 2009). 'One may say that World 3 is man-made only in its origin, and that once theories exist, they begin to have a life of their own: they produce previously invisible consequences, they produce new problems' (Popper and Eccles 1993).

Helmut Moritz deepens Popper's three worlds and their characteristics; the reality of World 3 in the field of mathematics and logic and their relation with exact thinking in the context of philosophy for scientists (Moritz 1995). Referring to cartography/GIS, Manfred Buchroithner, on the other hand, analyses potential multimedia geo-information in the overall system of Popper's Three-World-Model (Buchroithner 1997).

2.4.1 Popper's Three Worlds Theory and Cartography

Carla Lois (2009) analyses cartographic images and geographic imagery, and her claims have important impacts. She states that the presence of the map makes the absence of the object that the map represents visible. For instance, we cannot see the entire Earth because of its spherical form and size other than from space, but by means of maps (e.g. world atlases) we can view it well. In other words, the object is present (we are even standing on it) but it is not totally visible, we cannot see it as a whole.

> This means that there is a visual absence but not an absence of the object. But the object's representation is an image that not only pre-exists the object which, however, having been constituted in a permanent mediation, replaces it: the *representation builds the object.*[12] (emphasis added, translated by the authors from Lois 2009).

---

[11] Elements that belong to World 1 are for instance: stars and planets, atoms and molecules, tables and chairs, trees and animals, etc. To World 2 belong feelings, emotions, thoughts, pains, joys, wishes, etc. According to Gattei (2009); among others, words and prepositions; books and symphonies; laws; numbers and triangles (also problems, theories, and arguments) belong to World. 3. Indeed, elements of World 3 (e.g., a symphony) can have a physical presence in World 1 (a symphony recorded on a compact disc); it still belongs, however, to World 3.

[12] "Es decir, es una ausencia visual y no una ausencia del objeto. Pero la representación del objeto es una imagen que no sólo preexiste al objeto sino que, al constituirse en una mediación permanente, lo reemplaza: la representación construye al objeto." (Lois 2009, digital text: no pagination).

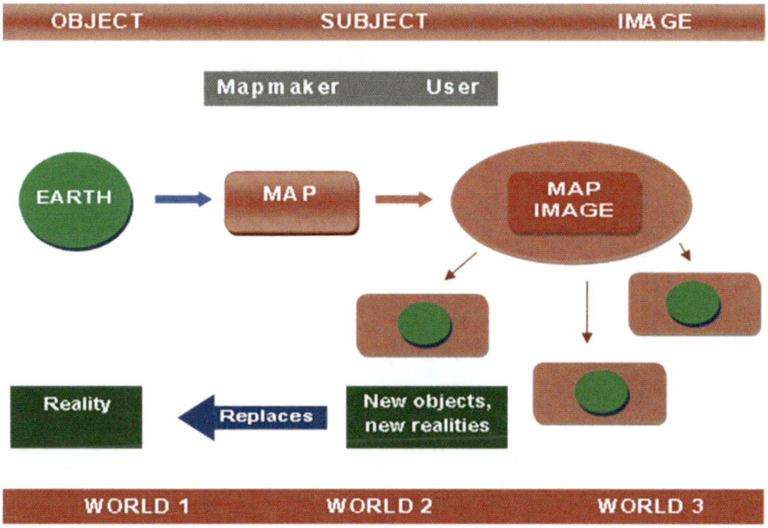

Fig. 2.1 Interaction between Popper's Three Worlds Model and the construction of new realities through maps in an ontological approach

If this quote is analysed in the light of Popper's interaction between World 2 and World 3, then it has important ontological and epistemological implication. A map depicts elements that belong to World 1 (e.g. the Earth). In this sense, the interpretation of the map's content (World 2) generates a map image[13] which belongs to World 3. The map as a device belongs to World 1 (artificial), but the map image belongs to World 3. Popper postulated that elements of World 3 acquire independent existence or autonomy. Thus, in the epistemological relationship between object-subject-image, the map image becomes the object. This means that the various map images that depict the objects, are transformed into objects themselves. In other words, the images are objects of reality and the images are themselves reality.

Figure 2.1 depicts the "Earth" as an object belonging to World 1 which is knowable through cartography. The subject is represented by the interaction between the mapmaker and the user through the map, and the latter belongs to World 2. The mapmaker designs and creates the map and the user reads and interprets it. Then the user generates a *map image* which belongs to World 3. From this mental image, new images are generated by the user. These images are regarded as *new objects* or *new realities* (belonging World 3) because they replace the phenomena of World 1.

Another aspect of the relationship between Popper's Three Worlds Model and cartography has to do with the term map (see Table 2.3). Cauvin et al. (2010) say

---

[13] In this case, we consider *map image* equivalent to "mental map" or "cognitive map".

Table 2.3 Parallelism between Popper's Three Worlds Model, products of cartography, and definitions and components of maps (after Azócar 2012)

Popper's three worlds model (*)		Cartography's product	Definition and component of map						
			(1)	(2)	(3)	(4)	(5)	(6)	(7)
World 3 (the products of the human mind)	Work of art and science (including technology)	Map mental image	X	X				X	
		Map model	X	X				X	
	Human language	Map language			X				
World 2 (the world of subjective experiences)							X		X
World 1 (the world of physical objects)		Map device				X			

(1–7) extracted from Cauvin, Escobar and Serradj (2010) (*) from Popper and Eccles (1993)
(1) Map is a geometric representation (of the planet; with relative positions; with non-spatial attributes)
(2) Map is a constructed model (reduction; selection; generalisation)
(3) Map is a graphical, iconic model using symbols (visual; audio; tactile; others)
(4) Map basis (permanent; temporary; virtual)
(5) Map at a given time within a context (historical; social; technological; scientific)
(6) Map with specific purposes (presenting and transferring information, providing locations; exploring patterns; revealing visible or invisible relations; exchange and consultation)
(7) Map involving choices (scientific; subjective; empirical)

that the name of the product of cartography is 'map' which is more familiar and more ancient than the discipline itself. They define the map concept and divide it into seven components (see Table 2.3). In comparison, the same table shows the content of World 3 by Popper and Eccles (1993) broken into 'work of art and science, and technology' and 'human language'. The cartographic products such as: map, mental image, and map model and map language (or cartographic language) fall into this classification. These elements belong to World 3 because of their cognitive nature.

Table 2.3 shows most of the criteria defined by Cauvin et al. (2010) within the table portion "Definition and components of map" which belong to World 3 (map, mental image and map model). The map belongs to these categories when they are regarded as a geometric representation, as a model by generalisation and selection, and for specific purposes (this task involves new information and decisions derived from map analysis). When a map is considered as a cartographic language (graphical and iconic model using symbols and signs), this implies that the map belongs to human language as a product of the human mind. If we look at Table 2.3, two criteria fall in the content of World 2: map at a given time within a context, and the map involving choices. These criteria are subjective because feelings and decisions are involved (for more details, there exists a wide body of literature authored by J.B. Harley). Finally, only one criterion belongs to World 1: the map considered as a permanent basis (i.e. material) implies independent of whether it is temporary or virtual. In this respect, Robinson et al. (1995) mention six major technological revolutions[14] starting at the time that cognitive images were first transcribed into tangible cartographic products (Robinson et al. 1995).

The criterion “Maps with specific purposes” requires a closer view. They are considered to belong to World 3 because they contain creations of new information by means of the map. This new information is as real as the previous one stemming from the set of input data. But the specific proposals which motivate this information quest can have different viewpoints (scientific, political, educational, etc.). As a result, these purposes can be subjective and experiential, and would therefore fall within the realm of World 2.

2.4.2 Three-Worlds Model of Popper and Multisensory Cartography

World 1 represents the “objective”, “real” world determined by physical, chemical (and biological) laws. *World 2*—Buchroithner (1997) is, for reasons obvious to cartographers but not discussed in more detail by him, inclined to call it “*World 3*”—is controlled by our mental and emotional processes, our sentiments and feelings: Based on the information perceived by our various senses, our thoughts and emotions (again, based on these sensations) we are building our own, “subjective” world or worldview. *World 3* finally consist of our scientific but also artistic activities describing the physical (“real”) world. They represent the underpinnings of our individual episteme (“*World 3*”).

Together with the epistemological approaches of the British philosopher, logician, mathematician, historian, and social critic Bertrand [Arthur William] Russel (18. 5. 1872 – 2. 2. 1970), with whose works Karl Popper dealt in a positive-critical way (cf. Russel 1948), Popper’s Three-Worlds Model forms an ideal background for both the information derivation from remote sensing imagery and for a cartographic information-transfer model, because it can be ideally combined with the concept of a spatial multimedia system (Buchroithner 1997).

Inspired by the idea to view, ‘in times of increasing interconnectedness and multimedia technologies, the science of cartography resp. geoinformatics not isolated any more but especially in the light of psychology and philosophy’, Buchroithner (1997: 31) crafted a German version of Fig. 2.2. There the author tried to realise a new view of cartography and to visualise the generation of our

[14] Mapping in the Western World has seen the following technical advances: manual, magnetic, mechanical, optical, photo-chemical, and electronic technologies (Robinson et al. 1995). All these technologies had impacts on the map as a tangible device belonging to Popper’s World 1.

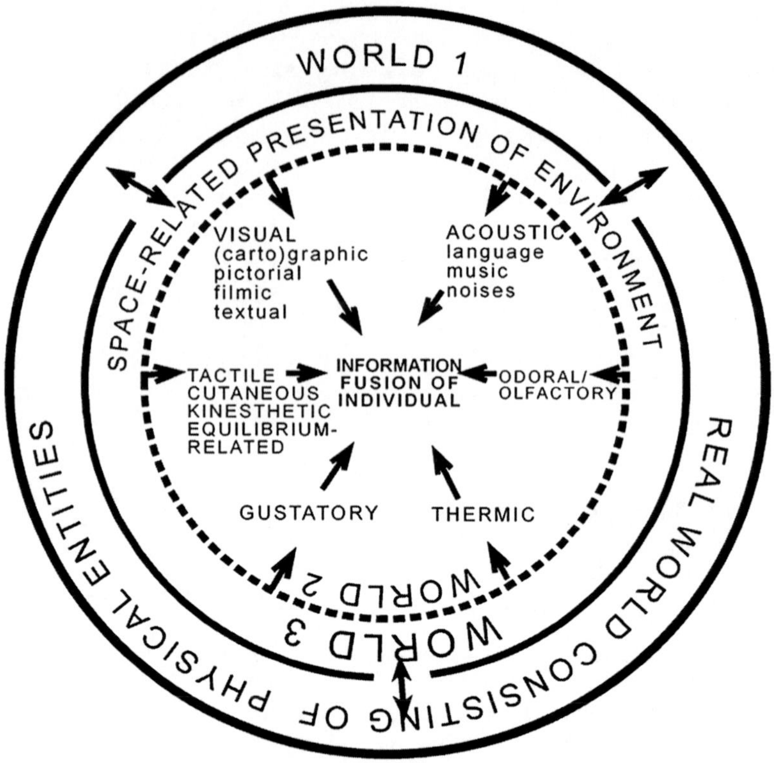

Fig. 2.2 Potential multimediate (multi-"sensory") spatial information within the overall system of the Three-Worlds Model (Popper and Eccles 1977) Furhter explanation see text below. (After Buchroitner 1997: 31)

subjectively perceived *World 2* out of information of *World 3*, which describes our physical environment, Popper's *World 1*. In the context of this figure it may be noted that, in addition to various other multi-"sensory" information transfers, since the end of the 20th century also sensations of smell have been conveyed together with visual and acoustic ones. In Fig. 2.2 the spatial information transmissions of common and well-tested types are represented by solid lines, sporadically materialised sensations by broken ones, and perceptions which are so far rather unlikely by dotted lines. A good example for the transmission of haptic, cutaneous or equilibrial sensations in a virtual three-dimensional world are flight simulators where also kinaesthetic sensations are, party in an exaggerated way, addressed (Buchroithner 1997).

Chapter 3
Paradigms in the History of Science

3.1 Introduction

The following pages will deal with the term *paradigm* as described by Thomas Samuel Kuhn (18 July 1922–17 June 1996) in the context of the history and epistemology of science. *Paradigm-shifts* or 'scientific revolutions' are analysed and their application in natural and social sciences is shown. Basically, the concept of paradigm implies several interpretations, but it generally includes a scientific community in a defined field which shares common aims and criteria during a certain period of time. These periods are called "normal science". The replacement of one paradigm by another is known as a "scientific revolution" or "crisis period". Kuhn proposes the *incommensurability concept*. This means that every paradigm has its own internal logic. Therefore the criteria of one paradigm cannot be applied to another one. As a consequence, each paradigm comprehends different world-views. The aim of this chapter is to study the reasons for a paradigm-shift as a basis for its analysis in cartography.

3.2 Kuhn's Paradigm Shift

In 1962, Thomas Kuhn published his seminal book *The Structure of Scientific Revolutions*[1] from which the term paradigm arose. According to Kuhn, paradigms are

> accepted examples of actual scientific practice – examples which include law, theory, application, and instrumentation together – that provide models from which spring particular coherent traditions of scientific research (Kuhn 1970: 11).

In other words, paradigms are universally recognized scientific achievements that for a time provide model problems and solutions to a community of practitioners (Kuhn 1970; viii). In the postscript to the 1970 edition of his book, Kuhn

---

[1] A complete outline of and study guide to the book *The Structure of Scientific Revolutions* by Thomas Kuhn was published by Frank Pajares (2005).

P. I. Azócar Fernández and M. F. Buchroithner, *Paradigms in Cartography*,

DOI: 10.1007/978-3-642-38893-4_3,

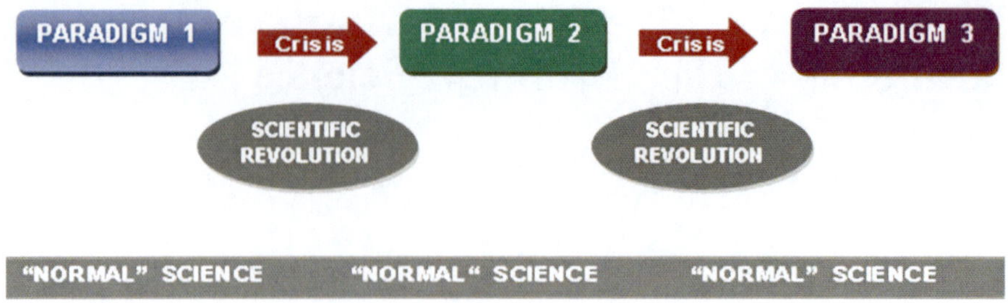

Fig. 3.1 Paradigms accompanied by scientific revolutions during periods of crisis and normal science. Adapted from Kuhn (1970)

formulates a generalisation to the term paradigm due to the difficulties derived from the multiple connotations[2]: "the paradigm is what is shared by members of a scientific community in particular" (Kuhn 1970) and by society. This is one of the most revolutionary insights in the history of science. Kuhn's theory of the nature of scientific revolution actually revolutionised the theory of knowledge in the second half of the 20th century and led to different social epistemologies. The study of the social dimension of science enables us to better understand the not rigorously-defined Kuhnian term (courtesy written communication by Zsolt Török, 2012).

The concept of paradigm is widely used in epistemology, and scientific knowledge is placed in *historical* and *social* contexts. Kuhn developed two essential components of the paradigm in his theory: a *disciplinary matrix*[3] and a *sociological component*. In other words, scientific thought does not progress via a linear accumulative, but on the contrary this thought is circular, *rupturistic*[4] and recaptures previous approaches or perspectives. The non-accumulative character of the scientific thougth permit to Kuhn propose the idea of revolution applied in sciences. In this way, Kuhn argues that periodic revolutions—called paradigm shifts—are immanent to science in a way in which the nature of the scientific inquiry within a particular field is abruptly transformed. This evolution of science includes "normal" science periods with a central paradigm followed by anomalies and revolutions ("crisis periods"). The arising of a new paradigm is an answer to the crisis.

Figure 3.1 depicts a period of normal science that follows a predominant paradigm (Paradigm 1) followed by a period of crisis when a scientific revolution

---

[2] Kuhn himself continuously changed the meaning of the term of paradigm and, according to his literary critics, never defined it exactly. He pointed out that "the term paradigm has been applied in at least twenty-two different manners" (p. ref.). Kuhn refers to the criticism made by Margaret Masterson in her article "The Nature of a Paradigm" (Masterson 1970).

[3] Disciplinary matrix: 'disciplinary' refers to the common posesion of the practitioners of a particular discipline; 'matrix' is composed of ordered elements of various sorts, each requiring further specification (Kuhn 1970: 182).

[4] Here *rupturistic* is considered to be synonymous of change, rupture, revolution.

Table 3.1 Some paradigm shifts in natural sciences, according to Thomas Kuhn

Field of study	Replaced paradigm (world view)	New paradigm (world view)
Physics (Astronomy)	Ptolomaic cosmology *	Copernican cosmology *
Physics (Mechanics)	Aristotelian Mechanics	Classical Mechanics
Physics (Mechanics)	Classical Mechanics	Quantum Mechanics
Physics	Newtonian Physics	Einsteinian Relativism
Chemistry	Phlogiston Theory *	Chemical Reactions and Combustion Theory *
Biology	Lamarckism Approach	Natural Selection Theory
Geology (Earth sciences)	Continental Drift Theory	Plate Tectonic Theory

Paradigm shifts (indicated with *) are broadly described by Kuhn is his book "The Structure of Scientific Revolutions" (1970)

takes place. A new period of normal science then arises accompanied by a new paradigm (Paradigm 2) until another period of crisis comes. This crisis is resolved with the rising of yet another paradigm (Paradigm 3), and so on.

In summary, Kuhn characterised the term paradigm as a *scientific tradition* involving theory, textbook problems and solutions, methodological apparatuses, and philosophy of science. Therefore, paradigms govern normal science. It is a scientist's task to apply the paradigm to the solution of puzzles.[5] A failure to solve these puzzles is the fault of the scientist, not of the paradigm. Nevertheless, persistent failure makes a puzzle an anomaly and threatens a revolution, which may end the paradigm's hegemony. A revolution is a fundamental change in the way of thinking about or the visualisation of something. In this case it is a *change* of paradigm.

Through the concept of paradigms, it is possible to explain the historical evolution of both physical and social sciences because they are defined by concepts, methods, theories, approaches, and problems. Table 3.1 shows some examples of paradigm shifts in natural sciences such as physics, chemistry, and biology. Both ancient and new paradigms are also considered to be *world views*, i.e., the philosophical and epistemological scope of viewing the world or reality.

For all conceptions, or world views, listed in the left column of Table 3.1, with the exception of geology, the old paradigms were replaced by new ones. In geology, the plate tectonic theory was a better explanation than, but complemented the former continental drift theory (Kuhn 1970). In other words, continental drift theory was not rejected, but it was during the 1960s and 1970s integrated into the new plate tectonic theory. Examples of this are more common in natural sciences than in social sciences. Some paradigms in natural sciences were developed and explained because Thomas Kuhn was initially a physicist and later became a theoretician of the history of sciences.

---

[5] Normal science is puzzle-solving. Doing research is essentially like solving a puzzle. Puzzles have rules and puzzles generally have a predetermined solution (Kuhn 1970).

On the other hand, Kuhn considers scientific knowledge to be *social* (Reyes 2005). This implies that scientists can be logically assigned to content-related disciplines or "invisible colleges" that are not consciously apparent to their respective members (Hall 2006, digital text: no pagination):

> Each college tacitly shares (a) a theory-laden vocabulary (based on implicit connotations as well as explicit definitions), (b) an unspoken set of examples of what the discipline believes to represent "good science", and (c) a lot of other uncritically held assumptions about their discipline inherited from their education as scientist (Hall 2006, digital text: no pagination).

Consequently, a scientific community cannot practice their trade without a set of *received beliefs*. According to Kuhn these beliefs form the foundation of the educational initiation that prepares and licenses the student for professional practice. The nature of a "rigorous and rigid" preparation helps to ensure that the received beliefs exert a "deep hold" on the student's mind (Pajares 2005, digital text: no pagination).

Kuhn also developed the concept of *incommensurability* in the framework of studying a scientific revolution, if there ocurred a progression from an earlier paradigm (disciplinary matrix)[6] to a newer one.

> Scientific revolutions may occur when new observations can no longer be adequately explained within an existing paradigm (the observations are anomalous). In some cases the anomalies can only be accommodated in theory based on new exemplars, models and/or symbolic generalizations. These changes often require new vocabulary and often alter the meaning and connotations of existing vocabulary. Even where the same words are used within each of the paradigms, there is often no longer a direct logical correspondence in their meanings. In other words, the world view (created by symbolic generalizations, models, exemplars and their associated theory-laden vocabulary) held by practitioners of one paradigm is logically *incommensurable* with that held by the alternative paradigm. Even though practitioners of both paradigms are looking at the same data, they see different worlds (emphasis added, Hall 2006, digital text: no pagination).

In Kuhn's theory, a new paradigm is *incompatible* with the paradigm that it substitutes or replaces. Therefore, a relation of incommensurability is generated. It implies that every individual theory fixes a meaning to all of its terms in a holistic way. If a small change happens in the theory, then the meanings of all the terms may radically change.

---

[6] According to Kuhn the concept of *disciplinary matrix* includes four major components: 1. Symbolic Generalisations— deployed by authors without question or introspection and immediately understandable by groups*[1]; 2. Models—including those with heuristic*[2] and metaphysical assumptions that provide groups with preferred analogies or even with an ontology; 3. *Exemplars*—which are unquestionable and accepted concrete examples of how to solve a particular kind of problem; 4. Values—in the sense of providing a predictive or epistemic value (Hall 2006).

*[1] Here "group" is considered a specific scientific comunity

*[2] In Philosophy, the adjective *heuristic* is used when an entity *X* exists to enable the understanding or knowledge of some other entity *Y*.

Incompatibility or incommensurability becomes clear when physical magnitudes are compared. For instance, terms like mass, velocity, and energy have completely different meanings in Newtonian classical mechanics compared with Einsteinian relativity. In the Special Relativity Theory the mass-energy equivalence principle ($E = mc^2$) arises, which is not valid in classic mechanics (cf. Reichenbach 1957). Therefore, the concepts of mass, velocity, and energy have different meanings in each of the theories; or these concepts are incommensurable.

Another important aspect of the theory of paradigms corresponds to the revolution as changes of world view. For Kuhn "after a revolution scientists are responding to a different world" and that a "scientist with a new paradigm sees differently from the way he had seen before". He asserts that "when *paradigms change*, the *world itself changes* with them" (Kuhn 1970: 111, 115, emphasis added). In other words, scientists see new and different things during scientific revolutions when looking with familiar "instruments" at circumtances that they have looked at before:

> This difference in view resembles a *gestalt shift, a perceptual transformation*–what were ducks in the scientist's world before the revolution, are rabbits afterwards (original emphasis by Pajares 2005, original text: no pagination).

This transformation of the world view is proposed because it changes not only a set of theoretical laws and a set of examplary achievements, but the agreements until then shared by the scientific community. Such agreements are: (Romaniuk and Paillalef 2010, digital text: no pagination)

- ways of raising problems in a field
- ways of speaking about the world (lexical)
- ways of seeing the world (ontological assumptions)
- ways of knowing the world in a reliable way (gnoseological assumptions)
- types of work, rules, instruments, techniques (methodological and practical assumptions)[7]

This new manner of seeing the world involves philosophical and epistemological aspects of studying reality due to lexical, ontological, epistemological, and methodological assumptions. For these reasons, the theory of paradigms has impacted fields such as the philosophy, history, and sociology of science. This is the *sociological* component of the theory. One example of this change in vision is seen in the philosophy of science during the twentieth century. In the first half of the twentieth century, the Vienna Circle was a milestone in logical positivism or neopositivism. However, criticism of the concepts of this approach allowed the emergence of other concepts which were different and embedded in the postpositivistic approach.

---

[7] Translated by the authors from Romanuik and Paillalef (2010, digital text: no pagination).

Table 3.2 Classical paradigms in geography, geography types, and their main representatives during the modern period of Western geography (after Azócar 2012)

Paradigm (Validity period)	Generated geographies	Main Representatives
Positivism (18th c.–19th c.)	– Environmentalism and Geographical Determinism	F. Ratzel, E. Reclus, P. Kropotkin
Historicism (1st third 20th c.)	– Regional Geography – Landscape Geography	V. La Blache, A. Hettner, R. Hartshorne, S. Passarge, O. Slutter, C. Sauer, M. Sorre
Logical positivism (1950s–1960s)	– Quantitative Geography – Theoretical Geography	I. Burton, D. Harvey, R. Chorley, P. Haggett, W. Bunge
Phenomenology (1970s–1980s)	– Humanistic Geography – Idealistic Geography – Perception Geography – Spatio-temporal Geography	– Yi Fu Tuan – L. Guelke – D. Lowenthal, P. Gould, K. Lynch – T. Hagerstrand
Radicalism (1970s–1980s)	– Radical Geography	D. Harvey, W. Bunge, I. Lacoste, R. Peet
Current tendencies (1980s–1990s and 21th century)	– Post-modern geographies[a]	

[a] For more details about postmodern geographies see Ortega Valcárcel (2000).

After Kuhn, there are authors who continue with the diffusion and illustration of the paradigm notion within other disciplines (Saldivia 2010). For instance, in the social sciences, Guillermo Briones conceives the paradigm concept as

> a conception of the study object of a science, the general problems to be studied, the nature of its methods and techniques, the information needed and, finally, of the way of explanation, interpretation or comprehension of the results of the performed research (Saldivia 2010: 63).[8]

In other words, each paradigm has its own study objects, research aims, methodologies, approaches, and research results. Adittionally, this implies the sociological component of the paradigms. Kuhn's legacy is viewed by postmodern and poststructuralistic thinkers as having called into question the enterprise of science by demonstrating that scientific knowledge is dependent on the culture and historical circumstance of groups of scientists rather than on their adherence to a specific method, as has been the aim Vienna Circle in the logical positivism approach.

[8] 'Una concepción del objeto de estudio de una ciencia, de los problemas generales a estudiar, de la naturaleza de sus métodos y técnicas, de la información requerida y, finalmente, de la forma de explicar, interpretar o comprender los resultados de la investigación realizada'. Here, Saldivia (2010) is drawing on the work of Briones (1987).

Table 3.3 Essential features of positivism and historicism.Extracted and translated from Capel (1983)

Positivism	Historicism
Methodological Monism (unity of science and scientific method)	Contrast between nature and history
Scientific Reductionism or Naturalism	Affirmation of the specificity of the human sciences
Nomothetics	Idiography
Explanation	Comprehension
Scientific knowledge uses only human reasoning	Empathic knowledge is valued as well as the use of faculties like sensibility and intuition
Prediction	Inability to do predictions in human sciences
Non-historicality	Emphasis on historical development
Axiological Indifference	Appraisal
Importance of theory (*)	Inductive methods without previous theories
Empiricism	Idealism

(*) In the inductive methods, dominant in the positivism of the nineteenth century, as well as in the deductive ones of the middle of the twentieth century

3.3 Paradigms in Geography

The concept of paradigms can also be applied to the development of modern Geography in the nineteenth and twentieth centuries. Some specific paradigms have been consolidated based on their historical, philosophical, cultural, economic, and political contexts. Different geographies have arisen within these paradigms. Each one presents its own study objects, research aims, methods, and results (Hernández 1982; Gomez Mendoza et al. 1988; Capel 1998).

Table 3.2 summarises the geographic paradigms, generated geographies, and their main representatives or authors based on the contemporary literature about Western geographical thinking.

On the other hand, Horacio Capel attempts to explain the evolution of geographic thinking in the nineteenth and twentieth centuries by comparing positivist and historicist stances. These two great traditions of Western thinking have been seen as two irreconcilable and conflicting concepts (Capel 1983).

Table 3.3 summarises in a schematic manner the contrast between the two opposite paradigms: positivism and historicism. In this way, two opposing world views with respect to the philosophical, epistemological, and methodological aspects of the knowledge of sciences are presented.

Capel (1983) poses the question: Are there paradigms in geography? He answers that since the beginning of the nineteenth century, the history of geography could be interpreted as a 'pendulum motion' between positivist and historicist stances. During the development of geographical thinking, this pendulum movement has two phases:

> The former would have its origin in the central period of the Enlightenment, … and it would be dominant in the middle of the nineteenth century and again, one century later, with the triumph of the Quantitative Revolution. The latter would impregnate the

> Romanticism of the beginning of the nineteenth century, the anti-positivism movement of the turn-of-the-century historicism, and of the first third of the twentieth century, and appears again in the reaction against to the neopositivism of the 'critical' and 'radical' geographies[9] (Translated by the authors from Capel 1983, digital text: no pagination).

This statement implies that the diversity of paradigms (as shown for example in Table 3.2) or geographic traditions[10] established during the development of the modern period of geography can be reduced to this pendulum movement which fluctuates between these two tendencies—positivism and historicism.

On the other hand, it is important to emphasise two other significant aspects of the theory of paradigms within the scientific context. First, although Kuhn explored the ideas of paradigms and incommensurability primarily in the temporal process of the change from one paradigm to another (Hall 2006), two paradigms can survive side-by-side at the same time, with direct consequences for the communication between holders of the different paradigms. Second, regarding the discussion about the relation between science and discipline, Kuhn's analyses applied in the scientific communities have also to be considered in general disciplines, whether they are scientific or not (Garcia-Sierra 1999).

It implies that the concept of paradigm can also be used for all those disciplines which are not regarded *strictly* scientific. Therefore, Garcia-Sierra (1999) proposes to use the term 'disciplinary community' rather than 'scientific community'.

---

[9] Las primeras tendrían su origen en el período central de la Ilustración [...] y dominaría a mediados del siglo XIX y, otra vez, un siglo más tarde con el triunfo de la revolución cuantitativa. Las segundas impregnarían el romanticismo de principios del siglo XIX, la reacción antipositivista del Historicismo finisecular y del primer tercio del siglo XX, y aparecen nuevamente hoy en la reacción frente al neopositivismo de las geografías "críticas" y "radicales".

[10] "Geographic traditions" are: physical, chorological, landscape, ecological, spatial, social, and socio-spatial (Capel 1983).

Chapter 4
Tendencies in Contemporary Cartography

4.1 Introduction

In the second half of twentieth century, major tendencies and perspectives arose during the formal and academic development of cartography and mapping. The analysis of theoretical trends in cartography goes back to Arthur Robinson's Ph.D. Thesis (1955). He was, however, not the first author setting the trend of *representational* cartography or a representative of modern cartography asserting that maps are objective, value-free representations. Robinson himself refers to Max Eckert whose monumental work *Kartenwissenschaft* (*Map Science*, 1921–1925) has to be considerd the manifesto of a new discipline (courtesy written communication by Zsolt Török 2012).

To this end, several authors and researchers have labelled these changes in the discipline with different terms: tendencies, trends, shifts, perspectives, approaches, paradigms, paradigm shifts, etc. In our book, we consider the changes to be mainly based upon the Western cartographic literature that show those characteristics that are pointed out in Thomas Kuhn's writing about the paradigm concept (Kuhn 1970). We take into account that surreptitiously these changes include the epistemological and philosophical bases, visions, and perspectives within applied scientific contexts, methods, and technologies, and their social context.

The tendencies were identified analysing the works of the following authors in chronological order:

1. Raul Ramirez discusses traditional and modern components in theoretical cartography.
2. Daniel Sui and James Holt present three major paradigms referring to map conception. Along the same line, Michael Peterson mentioned some paradigms associated with cartographic research and the Internet. *Cartographic Communication Model*, *Analytical Cartography,* and *Maps and the Internet* are presented as important tendencies in the Peterson's historical account.
3. *Cartographic Visualisation,* internationally established by Alan MacEachren, is considered to be an innovation in comparison with traditional cartographic communication.

P. I. Azócar Fernández and M. F. Buchroithner, *Paradigms in Cartography*,

DOI: 10.1007/978-3-642-38893-4_4,

4. In the context of *cyberspace* and digital cartography, the scope of view David Taylor makes as a basis of the *cybercartography* paradigm are presented.
5. Ferjan Ormeling's historical analysis of contemporary cartography in which he identifies some paradigmatic changes is also set forth.
6. The contribution of Colette Cauvin, Francisco Escobar, and Aziz Serradj about the main milestones in the history of thematic cartography during the last sixty years is analysed.

Discussions of specific works of these authors [or anything else, just so the reader knows what's ahead] follow in this chapter.

4.2 Traditional and Modern Components in Theoretical Cartography

In his book draft *Theoretical Cartography* Raul Ramirez highlights a difference between theoretical and applied cartography, in the context of modern cartography (Ramirez 2004). For Ramirez, there exist three major trends within theoretical cartography: the theory of *cartographic language*, the theory of *cartographic modelling*, and the theory of *cartographic communication*. They intend to explain the origin of cartography as a science and comprise the *classical theory of cartography* (original emphases by Ramirez 2004: 1).

The above statement mentions the rise of three *traditional components* or directions of theoretical cartography: *linguistics*, *modelling*, and *communication*. These traditional components will be treated in the following section in more detail.

4.2.1 Cartographic Language, Modelling, and Communication

Cartographic Language assumes that maps have a language that corresponds to a natural language used to express spatially related elements. Thus, cartographic language is composed of four elements: alphabet, grammar, reading, and writing. In the 1970s its main representatives were Lech Ratajski, Joel Morrison, Aleksandr Vasmut, and Jacques Bertin (Ramirez 2004). The cartographic language emphasizes that the language of the map has both an *alphabet* and a *grammar*. Alphabet had already been assumed through graphic elements (for instance, line, point, area, and volume). Grammar implies the combination of the mentioned elements. For more details about cartographic language or *cartosemiotics,* see Wolodtschenko

(2001, 2003, 2009, 2011) as well as Hruby and Wolodtschenko (2010) and *meta-cartosemiotics*[1] (cf. Wolodtschenko 2008).

Cartographic Modelling assumes that maps are models of the phenomena which are spatially represented. Cartography is considered to be a model of reality (i.e. especially of the physical world). This modelling is composed of four elements: generation of the model, analysis, interpretation, and evaluation (testing). In the 1980s its main exponents were Christopher Board, Konstantin Salishchev, and Ján Pravda (Ramirez 2004). This trend is based on the concept that a map is a *representational* and *conceptual model* of the real world, that there is a set of laws in maps (scale, presentation, and generalisation), and that maps are mathematically precise structures.

Cartographic communication considers maps as a means of communication. It has its origin in the need to demonstrate that in the design and production of maps, the opinion of users must be considered, because every model of communication is composed of a source, a message, and a receiver. This approach bears two elements: a *model of communication* and *theory of communication* (the latter one as a frame of reference). The major authors during the 1970s whose writings are representative of cartographic communication, were Anton Kolacny, Lech Ratajski, Joel Morrison, Christopher Board, and Arthur Robinson and Barbara Petchenik (Kanakubo 1990). This trend is based on the communication model of Claude Shannon derived from mathematics or signal-processing theory respectively (Shannon 1948).[2] This tendency will be analysed in greater detail later in this section.

4.2.2 Geo-Spatial Data Manipulation, Processing and Visualisation

Raul Ramirez (2004) also identifies three *modern components* in cartography which are directly related to manipulation, processing, and visualisation of geo-data. In the following brief descriptions of those components will be given.

Geo-Spatial Data Manipulation examines the representation of the terrain including the analytical study of cartographic documents, cartographic projections and their distortions, cartographic generalisation, cartographic transformation, and the quality of spatial data.

Geo-Spatial Data Processing analyses spatial information including the conceptual study of the topological representation of data, which is one of the

---

[1] Metacartosemiotics is 'a new concept that implies various changes in the range of theoretical cartography. These changes are both methodical and methodological kind and allow us to outline an evolutionary geo-communicational framework of cartography theory' (Wolodtschenko 2008, digital text: no pagination).

[2] The components of this model are: an information source, message, transmitter, signal, carrier or channel, noise (secondary signal that obscure or confuse the signal carrier), receiver, and destination. This model was adopted in cartography to explain how maps work.

structures used for analytical purposes. It includes these conceptual points of view: digital mapping, and geographic and land information systems. These systems are used for the acquisition, storage, management, and analysis of spatial data. Authors like Michael Goodchild and Martien Molenaar may be mentioned as early representatives of this view (Ramirez 2004).

Geo-Spatial Data Visualisation includes the conceptual study of all processes and forms of spatial data display, including the results of analysis and planning. It comprehends conventional methods, which include 2-dimensional (2-D) representations on traditional paper and graphic computer screen representations, and alternative methods (e.g. true three-dimensional (3-D) techniques like holography (Buchroithner 2012). The latter techniques provide means for generating truly 3-D static and dynamic displays supported by current multi-media technologies. During the 1990s, Alan MacEachren and David DiBiase were forerunners of theoretical research in this field (Slocum et al. 2007).

Ramirez further developed each component, especially those that were theoretical, supported by graphic diagrams. He described the major contributions to the development of the theoretical components in the light of cartographic theory (Ramirez 2004). Also in the 1980s, in the context of revolutionary changes in cartography, Joel Morrison defined four sets of processes compatible with those of Ramirez. These are data capture, data manipulation, data visualisation, and cartographic products (cf. Morrison 1986).

To sum up, it can be stated that the main contribution of Ramirez is the systematisation of the different tendencies developed during the second half of the twentieth century. His differentiation of the so-called *traditional components* (linguistic, modelling, and communication views) that were maintained through time is outstanding. His classification of geo-spatial data treatment—manipulation, processing, and visualisation—is proving to be valid until the publication of this book (2013), and especially true for the visualisation. It is also useful in the context of new information technology and mass-media in which images are playing an important role (cf. Pápay 2005a, b, c). The manipulation and processing of geospatial data in the context of geo-information technology is important because of the increasing amount of this type of data from diverse sources, such as remote sensing and global positioning systems.

4.3 Cartographic Research Paradigm and Research Focus

Through their study of the visualisation and analysis of public-health data, Daniel Sui and James Holt argued that paradigms should be engaged in order to better understand cartograms as a type of thematic map (in contrast to choropleth maps). They mention three major cartographic research paradigms, which have been identified in the cartographic literature of the post-World War II period. There are three different conceptualization of the *essence* of a map: (1) the map as image;

(2) the map as a model or computational tool; and (3) the map as intent or social construction' (emphasis added, Sui and Holt 2008: 5).

The authors also declare that the research focus can substantially differ, depending if it is on *map construction* or *map use*. In the following subsection, the three aforementioned paradigms and the varying emphases in research by Sui and Holt (2008) are discussed.

4.3.1 Map: Image, Model, Social Construction

'The map as an image' is a paradigm also known as the *communicative/cognitive tradition* (Sui and Holt 2008). It regards the map as an image emphasizing

> …the appropriate look and effective design of maps. In terms of map construction, this paradigm focuses on the design and visual symbols, the use of colour, the graphical hierarchy …; in terms of map use, it stresses map reading, visualisation, and communication (Sui and Holt 2008: 5).

Regarding this issue, Daniel Montello wrote a historiographical review of cognitive map-design research in the twentieth century, considering map perception, map cognition, and communication models under theoretical and empirical approaches (Montello 2002). He considered Arthur Robinson's book *The Look of Maps* (Robinson 1952) as an articulator of cognitive map-design research in several countries.

> On the other hand, 'the map as a model' is another paradigm known in an analytical tradition in which maps are conceived as a tool for analysis and modelling: 'In terms of map construction, it emphasizes data structure and algorithm development …; in terms of map use, this tradition often stresses analytical modelling and hypothesis testing of the clustering patterns of the phenomena being mapped' (Sui and Holt 2008: 5).

Furthermore, this is a paradigm framed within the field of analytical cartography, defined in the late 1960s by Waldo Tobler as an attempt of mathematical and quantitative analysis to solve concrete problems (Tobler 1976). Keith Clarke and John Cloud (2000) also continued the historical description of analytical cartography and its later application in parallel with Harold Moellering's work. For Moellering (2000, 2001a, b) this approach has grown into a broader and deeper scientific specialisation that includes the development and expansion of analytical/mathematical spatial theory and model building. Tobler's concept of cartographic transformation (Tobler 1979), Nyerges's deep and surface structure, and data levels (Nyerges 1980), and Moellering's real and virtual maps (Moellering 2000) have all been important contributions to the development of this paradigm in cartography.

Finally, 'the map as intent/social construction' is a paradigm framed in the critical tradition led by J.B. Harley (Crampton and Krygier 2006).

> Critical cartography represents a major epistemic break from the cognitive and analytical traditions; instead of conceptualizing the map as an objective, unproblematic device for communication, this paradigm reveals *the new nature of map* … (original emphasis by Sui and Holt 2008: 6).

The last phrase of the statement actually cites the name of the Harley's book (Harley 2001) in which a new dimension in the interpretation of maps is revealed. Sui and Holt, supported by authors like Denis Wood (1992), Jeremy Crampton (2001) and Harley (2001), point out that

> Because maps often make reality as much as they represent it …, mapping is, in fact, practices of power-knowledge. Many seemingly neutral maps express interests that are often hidden; thus, embedded in all maps is a set of power relation (Sui and Holt 2008: 6).

According to Crampton (2010), critical cartography and Geographic Information Systems (GIS) question the politics of mapping,[3] what kinds of people and objects are formed through mapping.

> Maps produce knowledge in specific ways and with specific categories that then have effect (i.e., they deploy power) (Crampton 2010: 9).[4]

In terms of map construction, this vision emphasizes the inherent distortion and biases of all maps, their power relation and ethical consideration. In terms of map use, 'the potential propaganda nature of maps for political hegemony and control (Crampton 2010)' should be recognised. This sentence explains the reasons that led Sui and Holt to employ the term 'map as intent/social construction' (Sui and Holt 2008). Here the term intent is synonymous to 'purpose'. In other words, the purpose of maps is to deploy power-knowledge. This topic will be discussed in greater detail in the next chapter.

4.4 Paradigms in Cartography: Cartographic Research and the Internet

In developing propositions about paradigms in cartography, Michael Peterson drew on some of Kuhn's arguments about the typical pattern of a mature science or discipline that has successive transitions from one paradigm to another through a process of revolution or change (Kuhn 1962). He also considered the characteristics of the paradigms pointed out by Kuhn. Further, Peterson argued that the concepts of paradigm and paradigm-shift are related to the rapid changes in cartography produced by the introduction of the Internet (Peterson 2002).

---

3 A critical politics of cartography is a problematization, a struggle, an ethics and, a technology (Crampton 2010).

4 This statement includes an important epistemological issue which was analysed in the previous chapter.

Regarding paradigms in cartography, Peterson, drawing on Kuhn, stated that 'a paradigm is a common core of beliefs about what represents a valid area of research' (Peterson 2002: 2), because research is guided by paradigms. In this way he identifies four paradigms within cartography: cartographic communication, analytical cartography, cartographic visualisation,[5] and power of maps.[6]

4.4.1 Cartographic Communication Model

Peterson (2002) states as a paradigm that cartographers began to view the cartographic communication process as a series of steps derived from communication theory. The map is considered to be a transmitter of messages encoded by means of a graphical language. This message is then decoded by the reader. Peterson proposes *cartographic communication model* which shows several stages between the cartographer and the map user.

A key aspect in his view is that maps are composed of elements that communicate information. Thus, improving the design of these elements through scientific methods would improve the communication potential of the map (Peterson 2002). In this sense, cartographers drew theories and methodologies from psychology, psycho-physics, and cognition. Peterson mentions that the interest in research related to cartographic communication faded during the 1990s, with the introduction of new tools like personal computers. Nevertheless, for Peterson,

> The research direction had a long-term influence on cartography because it had led to an increased understanding and interest in the discipline about communication with maps (Peterson 2002: 3).

This statement is in agreement with Jeremy Crampton when he describes 'how mapping became scientific' (Crampton 2010). The scientific status of maps and cartography was significantly highlighted by Arthur Robinson's experience and career, especially in the Office of Strategic Services (OSS) and at the University of Madison-Wisconsin.[7] Crampton also details how cartography became a scientific discipline (Crampton 2010). These events occurred during the validation of the cartographic communication paradigm, of which the *map communication model* (MCM) was the most important contribution.

---

[5] *Cartographic Visualisation* will still be further detailed later in this section.

[6] The *Power of Map* will be further analysed in the section "Critical Cartography in the Context of Post-Modernism".

[7] Arthur H. Robinson (1915–2004), American geographer and cartographer, was a prolific writer and influential philosopher on cartography. Generations of cartographers were influenced by his textbooks *The Look of Maps* (Robinson 1952), *Elements of Cartography* (1953) and *The Nature of Maps* (1976) (some of them with co-authors). According to MacEachren (1995), Robinson initiated a more 'objective approach to map symbolisation and design based on testing the effectiveness of alternatives'.

Several graphical models about cartographic communication by classical authors such as Anton Kolacny, Joel Morrison, Lech Ratajski, Christopher Board, Wiktor Grygorenko, Arthur Robinson and Barbara Petchenik have been compiled by Ulrich Freitag (2001), Raul Ramirez (2004), and Mirjanka Lechthaler (Lechthaler 2010), among others. These authors detail the elements that were a part of the cartographic communication process. Drawing upon his earlier paper from 1967, the year Jacques Bertin's Sémiologie Graphique was published, Christopher Board, in a comprehensive 35-pages article, made the sustained attempt to establish a set of theoretical principles that validate the place of cartography in the realm of scientific disciplines and to establish cartographic communication as a paradigm (Board 1967, 1972). Going back to the first concepts of Arthur Robinson he underpinned his construct with various models by European and American authors focussing on cartosemiotic aspects.

4.4.2 Analytical Cartography

The analytical cartographic approach focuses on the *transformations of information* inherent to cartographic procedures (Peterson 2002). Thus, 'it was in contradiction to the communication school because the *procedures* were more central than the map product or its use' (emphasis added, Peterson 2002: 3).

This statement describes a change in comparison to the map communication model of cartographic communication. Emphasis is put on the application of mathematical models and the subsequent application of technology. This new approach, which led to a paradigmatic shift, can be understood when Kitchin and his co-authors point out that analytical cartography emerged in the early 1970s 'offering a purely mathematical way of knowing the world …' (Kitchin et al. 2009: 6)[8]

Harold Moellering's work shows the origin and nature of analytical cartography and W. Franklin's oeuvre, its practical applications (Peterson 2002). Peterson contends that

> The goals of analytical cartography, which could be as mechanistic as improving the efficiency of a certain algorithm, were no match for the broader and more noble goal of improving the map as a form of communication (Peterson 2002: 3).

For Peterson, the above statement indicates that interest in the area of analytical cartography research remains limited. However, after a deeper analysis based on Kuhn's theory of paradigm (Kuhn1970), his statement can also be interpreted in a different way: If this approach has not prospered (as a normal process of scientific development), this could be an indication that analytical cartography correspond to the basis of a paradigmatic shift. In fact, several authors

[8] This statement made by the authors implies that, in Kuhnian terms, analytical cartography proposes a new viewpoint (mathematical model) in comparison to traditional cartography (communication model).

concur in that analytical cartography was the foundation for the emergence of the current geographic information science (among others Ormeling 2007, 2010; Moellering 2000, 2001a, b, 2012).

4.4.3 Maps and the Internet

It was shown in the early 2000s that research in cartography will include a new area due to the potential of the Internet as a major technological medium for cartography, especially concerning map distribution. For Peterson '…an individual map distributed through the Internet may not by itself communicate as much as the corresponding map on paper… [however,] the Internet makes it possible to distribute the map to many more people' (Peterson 2002: 5). In other words, map communication diffuses to more individuals through the Internet.

Peterson deems that similar to Kuhn's paradigm shift,'…the Internet has introduced a rapid, discontinuous change in cartography'. Thus, all the paradigms mentioned to this point, or essentially everything that was known about maps and their construction, could be replaced and lose their validity (Peterson 2002). This statement needs to be discussed. Peterson considers the Internet as a paradigm-shift in cartography. Nevertheless, he mentions another four paradigms in the discipline during the last half of the twentieth century that will be impacted by this new paradigmatic shift. The aforementioned statement seems, however, contradictics with Kuhn's statements about the development of paradigms.

Peterson's claim about new technological aspects of mapping is significant because it may generate important changes in cartography's theoretical, epistemological, and philosophical bases. Technology is an important aspect in the production of knowledge (cf. Azócar and Buchroithner 2009; Buchroithner and Azócar 2011). Indeed, as already mentioned, David Taylor proposed *Cybercatography* as a new theoretical construct, which 'offers an unprecedented opportunity for deeply rethinking the way we design, produce, disseminate and use maps on the Internet' (GCRC 2008, digital text: no pagination).

4.4.4 Trust in Map Information

Since the advent of Internet-based geoinformation, trust in map information has been playing an increasing role, particularly in the context of man–machine interaction (MMI) or human–computer interaction (HCI) and of Web Design. This goes hand in hand with the growing wish for the personalisation of maps, simply due to the fact that the user wants to trust in the depicted geodata. This requirement is best fulfilled if she/he her/himself gathers the data (cf. Buchroithner 2011: 24, 25, 31). Investigations into the trust of map information and into the various drivers influencing user confidence (depending on the respective situation) represent an upcoming field in times of various Internet platforms, social media, and personalized maps (cf. Schiewe and Schweer 2013 cum lit.), which will certainly have a feed-back onto map data acquisition and map design.

4.5 Cartographic Representation and Visualisation

In his book *How Maps Work*, Alan MacEachren proposed a fundamental dichotomy in approaches to the study of how maps work. First, he considered issues in the private realm of map percipients with emphasis on the perceptual and cognitive processing of sensory information. Second, he balances this approach with an analysis of map semiotics on functional and lexical grounds and an analysis of the public realm of maps. Finally, he concludes how this advocated, integrated perspective might be applied to an emerging area of cartographic concern (in the early 1990s): how maps work as visualisation tools. In other words, the author shows different types of maps which are used as applications in geographic visualisation (MacEachren 1995).

According to MacEachren, a representational view of cartography has two levels of analysis. The first one is the *private/perceptual – cognitive* level where:

> Attention is directed to how human vision and cognition represent concepts about the world and the contents of a visually displayed map, i.e. how meaning is derived from maps. ... The private focus is particularly concerned with the process of vision as a hypothesis about what is seen and the role of conceptual categories and knowledge schemata in assigning meaning to the representation derived by vision (MacEachren 1995: 15–16).

The second level is of the *public/social,* where:

> Attention is directed to the ways in which symbols and maps represent, i.e., how maps are imbued with meaning. ... The public focus is concerned with developing logical systems for creating meaningful representations and understanding in a broader context how symbols acquire meaning at multiple levels (MacEachren 1995: 15–16).

These two levels summarise this new perspective in cartography in a scientific view: geo-visualisation through semiotic/cognitive approaches.

Additionally, Michael Peterson schematically shows the principles of cartographic visualisation through two emblematic figures: *visual thinking/visual communication*, developed by DiBiase in 1990 (Fig. 4.1), and the *visualisation cube*, developed by MacEachren in 1994 (Fig. 4.2). For DiBiase,

> Cartography is defined into two fundamental activities: visual thinking and visual communication. Visual thinking occurs in the private realm and consists of the activities of exploration and confirmation. Visual communication, the public realm of cartography, involves synthesis and presentation (Peterson 2002: 4).

MacEachren introduced the *cartography cube* or visualisation cube'...that added the two dimensions of human-map interaction and the presenting knowns/revealing unknowns. Visualisation ... takes place in the high interaction-private-

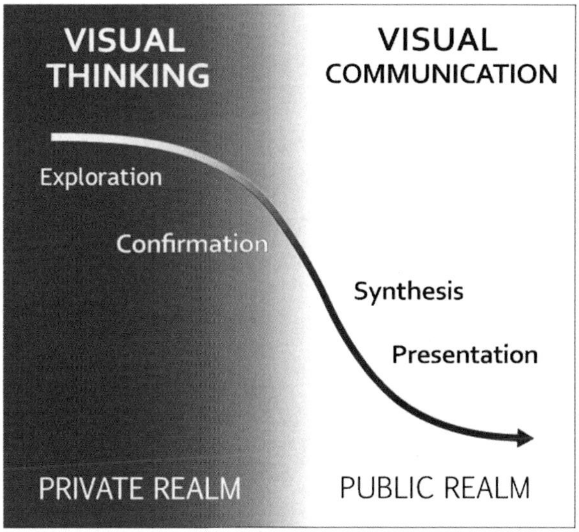

Fig. 4.1 Visual thinking and visual communication in cartography (after DiBiase 1990)

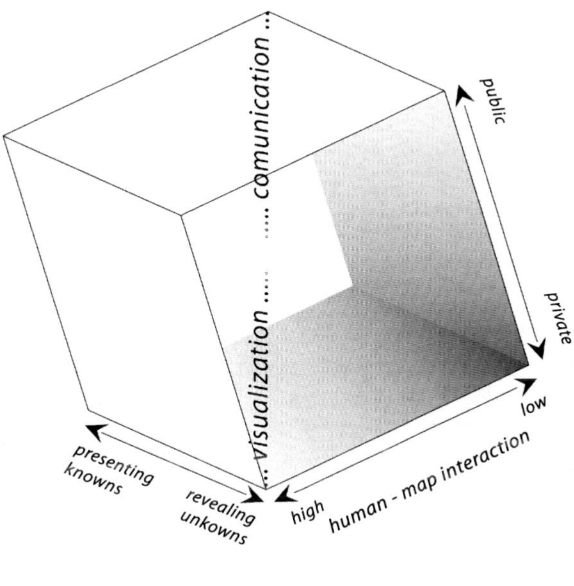

Fig. 4.2 Cartographic cube showing relationships between visualisation and communication (after MacEachren 1995)

revealing unknowns part of the cube' (Peterson 2002: 4). In other words, visualisation is emphasised when high human-man interaction occurs in the private realm with the goal of revealing unknowns, although communication may also occur.

Inversely, cartography takes place in the low-public-presenting known part of the cube. In other words, where low human-map interaction occurs in the public realm and known spatial information is presented, communication is emphasised.

Table 4.1 Integration between visualisation and communication approaches

	Visualisation	Communication
Visualisation approach	Visual thinking	Visual communication
Scientific research phase	Exploration/analysis	Synthesis/presentation
Map use (or realm)	Private	Public
Interaction human-map	High	Low
Data relation	Revealing unknown	Presenting known
Cartographic approach	Cartographic visualisation	Cartographic communication

Adapted from MacEachren (1994) and DiBiase (1990)

In this way, MacEachren (MacEachren 1995) depicts cartography as a cubic map-use space in which visualisation and communication stand in opposition. Table 4.1 show a synthesis MacEachren and DiBiase's contributions and the integration between visualisation and communication in the context of cartography. The participation of cartographic approaches in visualisation and communication are also shown.

Peterson considers the distinction between maps for presentation and maps for exploration artificial, because everyone who uses a map engages her/himself in the process of cartographic visualisation. The author draws from Unwin (1994) to state that the 'elite' aspect of cartographic visualisation emphasizes the use of graphics in the development of ideas, rather than in their presentation.

> The distinction between analysis and presentation with any type of graphic display is a tenuous one, especially in cartography. … Every map can be used for analysis, even maps on paper that are designed for *presentation* (original emphasis by Peterson 2002: 4).

A solution may, in part, be provided by Mirjanka Lechthaler's synoptic view of Menno-Jan Kraak's and Ferjan Ormeling's model of cartographic communication and the derived relationship between scientific visualisation and cartography (Kraak and Ormeling 2003).

> The cartographic presentation is a cognitive process which has to get the essence of spatial phenomenon, if it is represented adequately. … The objective of cartography … is to convey spatial information and their spatial relationships, the aims of communication, exploration and analysis (Lechthaler 2010: 387).

For a profound review of the theoretical bases of visualisation in science and of cartography in visualisation, the reader in kindly referred to DiBiase et al. (1992). Here, however, for the sake of better understanding the geo-visualisation approach, some further statements shall be presented.

Before the GIS era, paper maps and statistics were probably the most prominent tools available for researchers for studying their geospatial data. Since the early 1990s, people have access to large and powerful sets of computerised tools like spreadsheets, databases, and graphic tools (Kraak and Ormeling 2003). So, a comparison of the *on-screen approach* with the *traditional approach* not only reveals a difference in processing effort and time, but that the user can interacts with the map and the data behind it. Thus,

> The map should now be seen as an interface of geospatial data that can support productive information access and exploratory activities, while it retains its traditional role as a presentation device (Kraak and Ormeling 2003: 175).

The relationship between cartography and GIS on one hand and scientific visualisation[9] on the other has, from a cartographic perspective, resulted in a synthesis called geo-visualisation. Thus, according to MacEachren and Kraak (2001):

> Geovisualisation integrates approaches from scientific visualisation, (exploratory) cartography, image analysis, information visualisation,[10] exploratory data analysis (EDA) and geographic information systems (GIS) to provide theory, methods and tools for visual exploration, analysis, synthesis and presentation of geospatial data (any data having geospatial referencing) (Kraak and Ormeling 2003: 175).

In this way, the authors conclude that in a geovisualisation environment, maps are used to stimulate (visual) thinking about geospatial patterns, relationships and trends. So they might also offer different insights and would probably have more impact than traditional cartographic representation methods.

The above statements summarise MacEachren's thought of providing a new perspective to the scientific approach of cartography in comparison with the communication-cartographic paradigm. The semiotic-cognitive approach proposed by MacEachren (1995) is complemented by the early contributions by DiBiase, Krygier, and Reeves (1992), Kraak and Ormeling (1996) and also by MacEachren himself regarding scientific visualisation, geovisualisation, and their relationships to cartography and mapping.

In other words, geovisualisation represents a further development in cartography that takes advantage of the ability of modern computers to render changes to a map in real time, allowing users to adjust the mapped data at the same time.

Geovisualisation research flourished and spread during the 2000s, especially with respect to the cognitive approach. In fact, since 2009, cognitive issues in geographic information visualisation have been published in the scientific journal *Cartographica*. There, some topics of interest concerning the current fundamental empirical research and state-of-the-art evaluation methods within the interdisciplinary empirical research domain of geographic information visualisation and cognition have been shown (Fabrikant and Lobben 2009).

---

[9] Scientific visualisation focuses on the use of computer graphics to create visual images which aid in the understanding of complex, often massive numerical representation of scientific concepts or results. Also scientific visualisation has been defined as a multidisciplinary methodology and its specific goal is to act as a catalyst between scientific computation and scientific insight (Ed Ferguson 1991).

[10] Information visualisation is the interdisciplinary study of the visual representation of large-scale collection of non-numerical information, and the use of graphical techniques to help people understand and analyse data. In contrast to scientific visualisation it focuses on abstract data sets that do not have an inherent 2D or 3D geometrical structure.

Similarly, Menno-Jan Kraak reviews the development of cartography as seen by the International Cartography Association (ICA) Comission on Geovisualisation, from 1970 through the end of 2000. The most obvious tendencies are: cartography, computer cartography, geovisualisation, and visual analytics.[11] During this period, the amount and diversity of data increased tremenduously, last but not least, through the use of remote sensing-based information retrieval. This is also the period where developments in cartography began to be much more technology-driven (Kraak 2008).

During the years just before and after 2000, the *visual analytics* approach was translated into the geographic information science (GIScience) domain as *geovisual analytics* (cf. Kraak 2008). This field is about analytical reasoning and decision making. This new trend,[12] '... requires a multidisciplinary approach with strong analytical capabilities in which geovisualisation is a major thought-provoking tool' (Kraak 2008: 164). In other words, this shows the importance of cartographic representation and visualisation in the transition from *geovisualisation* towards *geovisual analytics*. Drawing from Kraak's paper the interplay between cartography, geovisualisation and geovisual analytics, using an extended definition of the latter term, is comprehensively treated by Schiewe (2013).

4.6 Cybercartography Paradigm

The concept of cybercartography was introduced by David Taylor in 1997 during the International Cartographic Conference-ICC in Sweden. For the author,

> The central argument made was that if cartography was to play a more important role in the information era, then a *new paradigm* was required (emphasis added, Taylor 2005: 1).

It implies that the importance of maps and mapping and the utility of cartography must be reasserted and demonstrated in the context of rapid advances in information and communication technology. Taylor sees the paradigm of cybercartography not as a break from past ideas and practice,

> ...but as an evolutionary and integrative process which incorporates important elements from the past, redefines others, and introduces new ideas and approaches to both cartographic practice and theory (Taylor 2005: 2).

For Taylor the dominant paradigm until the 1990s had been cartography as a science. Indeed when he analyses the technological background of the discipline, he notes:

---

[11] Visual analytics is based on the intuition that highly interactive and dynamic depictions of complex and multivariate databases amplify human capabilities for inference and decision making, as they facilitate cognitive tasks such as pattern recognition, imagination, association, and analytical reasoning (Thomas and Cook 2005).

[12] Here Kraak draws on the work of Andrieko et al. (2007).

> There has always been a strong formalist base to cartography as a discipline and computer technology has led to an emphasis on productive techniques to which GIS has added a strong emphasis on *positivism* (emphasis added, Taylor 1994: 53).

Thus, the emergence of this new paradigm has a strong scientific component, but the author sees:

> …cartography as both an *art and science*, and has a qualitative as well as a quantitative element (emphasis added, Taylor 2005: 3).

This joining of art and science in cartography is an important idea, because inside the paradigmatic scientific approach, there has traditionally been a gap between aesthetics and accuracy as objectives to be attained by cartographic products. In this sense, the paradigmatic aspect pointed out by Taylor implies a process of integration through the new technologies, which became massive during the early 1990s.

In summary, for Taylor cybercartography is

> The organization, presentation, analysis and communication of spatially referenced information on a wide variety of topics of interest and use to society in an interactive, dynamic, multimedia, multisensory and multidisciplinary format (GCRC 2008).

Similarly, Maria Del Carmen Reyes (2005) also saw the emergence of a new paradigm in cartography that 'is transforming the manner in which a map is conceived, produced, and used. There is currently a need to focus on *epistemological* and conceptual issues behind cybercartography, in order to develop a common language amongst researchers and a *theoretical framework* that results in the scientific advancement of this discipline (emphases added, Reyes 2005: 65).

This paradigmatic requirement is also evident for Elvia Martinez and Maria del Carmen Reyes (2005) when they draw on Kuhn's ideas and then relate them to David Taylor's statements. Kuhn considers scientific knowledge to be social. The concept of paradigm is also used by Kuhn 'to depict the network of convention and compromises, created among scientific communities in order to produce and legitimate scientific knowledge' (Martinez and Reyes 2005).

This train of thought appears in a similar way when Taylor argues that cybercartography is a new paradigm in cartography (Martinez and Reyes 2005). In cybercartography, there is a group of researchers that share an epistemological view and generate a new body of knowledge. Thus, this body of knowledge can be validated both theoretically and empirically.

At the time of this publication, the main products of cybercartography are *cybercartography atlases*. According to Taylor…

> A cybercartography atlas is a metaphor for all kinds of qualitative and quantitative information linked through their location[13] … Cybercartographic atlases transform cultural,

[13] For more detail about types of atlases and research in this field, see Geomatic and Cartographic Research Centre GCRC (2008).

> socio-economic and environmental data into interactive and multisensory narratives.... These atlases present several narratives of the same *reality*. People can be become active *creators* of map narratives, not just passive ones (original emphasis by Taylor 2009).

With cybercartographic atlases, as a specific product of cybercartography, there has been a shift from map user to map creator. In other words, cybercartographic *users* can become *creators*. Hence the cartographic process is democratised in new ways.[14] In this way, cartography belongs really neither to academia nor to the professional cartographers.

Finally, the underlying philosophy of the cybercartographic atlas can be formulated: 'Just as the map was a key navigational tool in the Age of Exploration, so the cybermap can provide an aid to navigation in the information era' (Taylor 2009).

This metaphor implies that technological changes have always had an impact throughout cartography. Here, the author mentions two significant historical events (i.e. the eras of exploration and information) in which maps, supported by technology, have generated and still can generate new knowledge again.

4.7 Paradigm Changes in Cartography

Ferjan Ormeling compares the general cartographic practices of the 1970s with those from then until 2010. He identifies several changes and their consequences for the future of cartography. In his article about the transformation of the map into a multifunctional signpost, he points out some paradigmatic changes in cartography (Ormeling 2010).

Briefly, the author restates several definitions of the term cartography through time. In 1820, the term merely encompassed the *production of maps* (map production). In the 1960s, cartography was defined as the *communication of spatial information*. This is a process subject to specific rules in the design of maps for a proper presentation of geographic information. In 1967, the application of Kolácný's model[15] (Kolácný 1969) in cartography provided a scientific approach to the transfer of information. Thus, psychophysiological research was applied in cartography. For this reason, as from [the developments of the 1980s, the term cartography has referred to the *production and use of maps* (Ormeling and Kraak 1987).

Ormeling notes that after 20 years of automation with computer-assisted cartography, it was not only possible to produce maps, but

[14] For processes of democratisation also see Ormeling (2007, 2010).

[15] According to Czech cartographer Anton Kolácný, the creation and utilisation of cartographic products are two components (cartographer's universe and user's universe) of an interrelated process in a *stimulus–response* model (Lechthaler 2010). This incorporates multiple feedback loops and interconnections in the previously simplified map communication model composed by the cartographer, map, and percipient (Crampton 2010). Therefore, this model became more complex with Kolácný's contribution.

> Once one had stored the spatial information needed to draw maps in the computer, one could also begin doing some calculation: determining area, measuring distances, and carrying out visibility analysis (Ormeling 2010: 7).

The author stresses that these works belong to analytical cartography and that the new methods of analysis gave birth to GIS.[16] Ormeling states that the arrival of digital geographic files led to a revolution in map production. In this way, 'Once the information was digitally stored in a file, one could easily visualise that which was needed for a certain purpose from that file' (Ormeling 2010: 7).

The above statement meant that for the first time in the history of cartography it was possible to separate the *storage function* from the *communication function* of the map (cf. Ormeling, 2007). The author exemplifies this important change with a nautical chart having a complete set of publication content, comparing it to a digital version for individual use. In other words, with digital files (containing geographic information) it is possible to display on screen only the information required for specific objectives.

By 1992, Ormeling had analysed core concepts in cartographic communication, when he pointed out two revolutions in cartography. In the first one, called the *communication revolution,* he sees maps as a means of spatial information transfer, in the second one, the *digital revolution,* as a separation of storage and display functions (Ormeling 2007).

This results in the following:

> This breakthrough changes the content of the term cartography once again: now cartography stands for *passing on spatial information to support decision making* (original emphasis by Ormeling 2010: 7).

In summary, despite mentioning the term paradigm only in the subtitle of his article, it is possible to infer from Ormeling's publications a paradigmatic shift in the development of cartography the first paradigm was map production. Then spatial (geographic) information via visualisation became a second paradigm: map production *and* map use. Meanwhile, analytical cartographic (geo-visualisation) and geographic information sciences were encompassed during the automation of processes. Finally a third paradigm is considered: spatial information to support decision making. The separation between storage function and the communication function was the great impulse for the emergence of this paradigm shift.

---

[16] For more information regarding analytical cartography and GIS the reader is referred to Moellering (2000), and for the relationships between cartography and geographic information systems see Cassettari et al. (1992) Grelot (1994) as well as Lee (1995).

4.8 Cartographic Trends and Paradigms Since 1950

In their studies regarding the development of cartography, Colette Cauvin, Francisco Escobar, and Aziz Serradj identified in their book entitled *Thematic Cartography and Transformations* (2010, Vol. 1) the general trends and paradigms that arose from 1950 to 2009. They divided the second half of the twentieth century and the first decade of the twenty-first century into three periods, pointing out some milestones, such as books or remarkable events, dominant paradigms or ideas, technological changes, and crosscutting current of thought, that will be discussed in this section (Cauvin et al. 2010).

The first period (approx. 1950–1975) was dominated by two paradigms or ideas: the map as a channel of communication and the rules of graphical semiology. Two major books are commonly associated with this period: *The Look of Maps* by Arthur Robinson (1952) and *Semiology of Graphics* by Jacques Bertin (1967), respectively. It ends with the Vienna Congress (1975) in which cartography was—once again—defined as science, and the value of theoretical cartography was highlighted. During this period, a fundamental change in cartography with regard to aspects of graphic concept, language, perception of signs, and the function of maps took place.

Three dominant paradigms or ideas were central to the second period (about 1975–1995): theory of symbolization and design, experimental and exploratory cartography and, inclusion of ethical and social aspects. One of the major technological changes in this period was the complete separation between the storage and representation of data, and the proliferation of GIS and multiple data processing. This meant that a temporary map (on the screen) substituted for a permanent (paper) map.

During this second period some trends revived, such as: graphical perception and cognition, in which the reader of a map should be stimulated by the map and no longer considered a simple recipient of its message. Later, the context in which maps are produced was also a conceptual trend in cartography. In this sense, the historical, political, and social context implicitly or explicitly interferes in map production. Another important trend is the exploratory use of maps. Visualisation in scientific computing allows maps to have a new role: as a useful means of obtaining spatial information for users.

Finally, the authors distinguish a third period (roughly 1995–2009), in which there arose a new paradigm for the twenty-first century: geovisualisation. In this period, two changes can be identified: the integration of cartography into GIS and the shift from the communication paradigm towards scientific visualisation. In this way the aforementioned visualisation paradigm allows the exploration of the information in a dynamic way, by means of the development of the man-computer interaction. The objective is not to obtain an 'optimal map' but an 'efficient map', based on the concepts of visual perception and spatial thinking.

The new elements in the context of geovisualisation are: multiple representations, dynamic cartography, animation, interactive maps, multimedia, hyper-maps,

Table 4.2 Main tendencies and changes in cartography and mapping during the second half of the twentieth century (After Azócar 2012)

Views on cartographic development	Tendencies and paradigm shifts of cartography and map conception
Traditional components and modern components (Ramirez 2004)	Cartographic language Cartographic modelling Cartographic communication Geo-spatial data manipulation Geo-spatial data processing Geo-spatial data visualisation
Cartographic research paradigm and research focus (Sui and Holt 2008)	The map as image The map as model The map as intent/social construction
Paradigm in cartography: cartographic research and internet (Peterson 2002)	Cartographic communication Analytical cartography Cartographic visualisation Power of maps Maps and Internet
Cartography: representation and visualisation (MacEachren, MacEachren 1995) (Kraak and Ormeling 1996, 2003)	Cartography as graphic communication Cartography as geo-visualisation
Cyber-cartography paradigm (Taylor 2005)	Traditional cartography Cyber-cartography
Paradigm changes in cartography (Ormeling 2007)	Production of maps Map production and map use Spatial information to support decision making
Cartographic trends and paradigms (Cauvin, Escobar and Serradj 2010)	The map as a channel of communication Rules of graphical semiology Theory of symbolisation and design Experimental and exploratory cartography Ethical and social aspects Geovisualisation

web maps, maps on demand. There exists also something similar to collaborative cartography, in which the reader is an active participant.[17]

Table 4.2 summarises the tendencies in cartography since the second half of the twentieth century. The listed authors have had a theoretical vision of cartography and mapping. The changes identified by them have a certain similarity with Kuhn's paradigm shifts (explicitly or implicitly) in that these changes are visions or perspectives that are different from each other. The different modes in which to conceive cartography (as a discipline, as science), maps (as models, devices, products), methods of analysis, and technologies, lead different study objects, objective or research, methods, and results.

[17] For more details see Cauvin et al. (2010).

4.9 Further Tendencies

4.9.1 VR Cartography

In 2005, Manfred Buchroithner introduced the term "*VR Cartography*" to the global cartographic community at the 22nd International Cartographic Conference in La Coruña, Spain, (Buchroithner 2005). In that particular case, the author referred to a combination of a classical, solid, relief model draped with texture derived from a satellite image and an interactive, virtual, flight view over the depicted mountain area. Asking 'How can the over whelming beauty of the landscape and all the habitat particularities be presented in an informative and challenging way?' Buchroithner chose the combination of observation of the solid terrain model (STM) and a helicopter pilot's view for an installation in an Alpine national park visitor centre.

In fact, simulating a flight over an Alpine landscape represented by the landscape model, the user has the possibility to enjoy a 45° downward view out of the cockpit on a 150 × 200 cm^2 panorama screen behind 200 × 300 cm physical relief model. Thus, operating the virtual helicopter by a joystick at a console in front of the landscape model, the user is able to combine the stereoscopic view of the down-scaled landscape (with the foot-point position of the helicopter indicated by a moving laser-light spot on the landscape model) and the 'pseudo-3D view' of the landscape in front of her/him. Interactively, name-tags of various points of interest in the form of billboards can be switched on. In addition, further explanatory texts and/or audio-information can be selected. By minimising the flying height to a distance of about 175 cm and slowing down the speed, a virtual walk-through can be experienced. Since the above mentioned simulated views are calculated 'on the fly', Buchroithner coined the expression *Interactive Real Time VR Cartography.*

4.9.2 KIS: Cartographic Information System—a (Short) Episode (?)

A development that took place around the first half of the 1990s and that—to the authors' knowledge—actually might have had its roots in the Vienna School of Cartography, shall be briefly mentioned: the *KIS* (*Kartographisches Informationssystem*). Fritz Kelnhofer (and his team) proposed his idea of *cartographic information systems* in a series of papers published between approx. 1990 and 1995 (Kelnhofer 1993, 1994, 1995, 1996; Kelnhofer and Lechthaler 1995). The author, drew on publications by Grünreich (1992), Ormeling (1996), and others, and based on profound theoretical deliberations, partly published in earlier notes related to this approach, such as "*Cartography as Basis of GIS*" (Kelnhofer 1993), A *KIS* makes use of the well-known differentiation between the primary object models and the secondary models of cartographic visualisation. Based on GIS databases, cartographic products can be generated by means of multifunctional application of spatial and attribute data in an elegant way. In his essential paper Kelnhofer (1996: 20, 22, 21) gives a comprehensive and theoretically well-reasoned description as well as a detailed workflow graphic of the functioning of a complex KIS, also setting forth the advantages of a KIS over a GIS in terms of the production of high-quality map products, be they digital or analogue.

Around the year 2000, the developments following these first methodical, structural, and implementation activities led to the generation of an optimised interactive *KIS* with a rather flexible modular structure (Brunner-Friedrich and Stadler 2006). Maybe, the whole epistemological genesis of cartography in the following years would have changed if the cartographic community would have put more emphases on *KIS* instead of *GIS* and on the final products, i.e. the maps, than on the data bases and methodologies behind them, i.e. the system architecture. In particular in the Anglo-American world, however, at this time the *GIS freakiness* was overruling the importance of the final outcomes, the *geovisualisations*. Thus, the strange situation occurred (which is unimaginable in other disciplines or sciences such as medicine), that, the terminology, the technologies, and methods ruled out the actual target or "desired object", the maps (cf. Buchroithner 2011: 24).

4.9.3 Cartography and Visualistics[18]

As repeatedly mentioned in their seminal book *Representation in Scientific Practice* by Michael Lynch and Steve Woolgar (1990), images serve to showcase scientific results. They do not only present visible phenomena but, also support the visualisation of the invisible. They play an essential role in the conveying of knowledge: "Only through scientific drawings, graphs, diagrammes, computer programmes, photographs, x-ray images, *cartographies* etc. scientific knowledge is brought into a stable, expressive and communicable form"[19] (Domman and Meier 1999).

In this sense, the Hungarian historian and cartographer Gyula Pápay, drawing upon the instructions of Claudius Ptolemy for map generation, denoted cartography as the oldest visualistic science ("Bildwissenschaft"; Pápay 2005c: 8). He further referred to the '*pictorial turn*', also called '*imagic turn*' or 'iconic turn' when wrote about the period when "images" became important in science.

---

[18] In German: *Bildwissenschaft* (*Image Science*). It has, however, to be mentioned at this point that—in contrast to cartography—in the visualistics or image science proper as well as in semiotics, the concept of models is, surprisingly enough, to the greatest extent dismissed.

[19] 'Durch wissenschaftliche Zeichnungen, Graphen, Diagramme, Computerprogramme, Photographien, Röntgenbilder, *Kartographien* etc. wird wissenschaftliches Wissen überhaupt erst in eine stabile, aussagkräftige und kommunizierbare Form gebracht.' (emphasis added' Dommann and Meier1999: 15).

There is broad agreement among cartographers that cartography can, due to its interdisciplinarity, be assigned to the spatial sciences, the geosciences, the environmental sciences, the information science, the communication sciences (cf. Azócar 2006a, b, 2007–2008; Azócar and Buchroithner 2009; Buchroithner 2011; Buchroithner and Azócar 2011), but also to the *Bildwissenschaften*, the "image sciences" (cf. Pápay 2009). In his 2009 publication Pápay not only dealt with the cartographic spaces in the overall system of space types. It is to his credit that he tried for the first time to determine the invariant properties of cartographic space from both a spatial-scientific *and* an image-scientific/visualistic view, thus breaking with traditional concepts (cf. also Pápay 2012).

Within the past decades, last not least due to the advent of high-resolution spaceborne imagery, map-like representations gained increasing importance in cartography. Traditionally, in contrast to maps, they are considered non-Euclidian. This, however, is a questionable differentiation, since they represent a combination of concrete and abstract spaces, whereby the spatial structure of concrete space, the *geo-space*, is exclusively represented topologically (Pápay 2009). Pápay referred to them further—he called them "remarkable"—as map-related products which exceed the topological idealisation by generalising the spatial relations but simultaneously depicting the generalised topology in a pictorial way. He introduced the term *Kartoiden* (*cartoids*) for them (Pápay 2005b: 292).

They represent a transition form to those geo-portrayals where abstract spaces such as social spaces, communication spaces, or cognitives spaces are visualised by concrete image spaces (cf. Picker et al. 2013). Although they are not "real maps" in a cartographer's understanding, they are also called "maps" (Pápay 2009: 170).

For all that, of course, the definition of *image* is crucial. Drawing upon the American semiotic philosopher Nelson Goodman (Goodman 1968), Umberto Eco called it a naive conception that *resemblance* is an essential property of an image (Eco 1975, 1985). With the "new" view of images they have been considered as a particular type of signs, thus offering the possibility to search and analyse an image according to syntactic, semantic, and pragmatic aspects (cf. Pápay 2005a, b, c), something which is in common with maps, too. The semiotic conception of images, which has already been initialised by the founders of semiotics, in particular by Charles S. Pierce, is nowadays state of the art. For the development of a *General Symbol Theory,* Nelson Goodman played an essential role and might be called its godfather. He discovered the singularity of pictorial sign systems due to their syntactic density. One of the big advantages of images over verbal signs is that the syntactic structure of images is dense and not *disjunct* as with character-signs. However, half-tones of hill-shading in a map, for instance, are *continuous* (cf. Nöth 2005; Pápay 2005a); so are many more areal and linear features in maps.

Further, following the famous Greek legend describing the drawing of a contour line of a cast shadow of a person as "the origin" of painting (cf. Athenagoras, Pliny the Elder, Quintilius), the shadow is considered a natural image, an *icon*, and the contour, a *symbol*. This concept has for long been applied in cartographic presentation. Basically, *topoi*, places or locations, are symbols, and their spatial relations are iconic. Flowing waters synthetically combine symbolism and iconicity: their graphic depiction is the result of a symbolic denotation, their course, on the other hand, is iconic. In this sense, cartographic *pictograms* have a double denotation through the superimposition of the iconic denotation and a symbolic one (Pápay 2005c: 17).

In a very basic sense, maps represent combinations of icons and symbols by portraying relief by means of hill-shading. Settlements may be represented by pictograms. In this case there exist both a primary and a secondary iconic denotation. The first one is superimposed by the explicit symbolic denotation, and the latter one is used for the rendition of spatial relations of the individual pictograms. Also, for the combination of all three elementary pictorial types—icon, pictorial symbol system, and individual pictorial symbol—maps serve as paramount examples for complex images, as already mentioned above: hill-shaded relief and pictogram-type settlement representations stand for that (cf. Pápay 2005c).

In summary, drawing on Papay (2005c: 18), it can be concluded that in cartography there exist three types of combinations of iconic and symbolic denotations: *the superimposing combination*, the *synthetic* or *hybrid combination* (cf. the rivers) and the *individual use of both denotation types side by side* in one and the same "image", i.e. map.

Principally, two modes of dealing with imagery can be distinguished: the *symbolic* and the *immersive mode* (Sachs-Hombach and Schirra 2000). Once the user knows that he is dealing with so-called *secondary images* (i.e. images which are beyond our visual perception like landscape paintings or photographs; cf. also Wittgenstein's and Popper's views) or *tertiary images* (images that represent an objectification of constructed images), she or he is in the *symbolic mode*. If, however, the illusion is so massive, that the actual secondary or tertiary images are taken for *primary images*, then one is acting in the *immersive mode*. This, again, has an impact on cartography because with photorealistic cartographic (i.e. geometrically defined and labelled) scenes in (auto)stereoscopic vision, or with augmented reality (cf. *VR Cartography* (see Sect. 4.9.1. and *Autostereoscopic True—3D Cartography—another new paradigm* (see Sect. 7.6) one is dealing with the realm of *immersion*.[20]

Pápay's approach of considering cartography—among others (see above)—as a *Bildwissenschaft* is by all means worth mentioning; maybe due to the fact that he only published in Hungarian and German, it was underestimated, at least in the Anglo-American world. The credit belongs to him that he dared to break with the *traditional view* that *maps are no images*. The increasing use of both cosmic and terrestrial imagery in modern geovisualisation may support this notable approach to maps and cartography. In his comprehensive exposition about "cartography as image- and spatial science" Pápay (2012) explained the historical reasons for the long-lasting aloofness of the cartographers towards the term "image" and examined

[20] Wikipedia defines *immersion* in the context of *virtual reality* as 'the state of consciousness where an immersant's awareness of physical self is diminished or lost by being surrounded in an engrossing total environment'.

image-science aspects of cartography in the context of space-theoretical aspects and presents, based on his previous publications, a valid and persuasive concept for the cartographic image space.

4.9.4 "Living Cartography": Technology-Driven Versus Design-Driven Cartography

In a remarkable book review of the North American Cartographic Information Society's recently published *Design Atlas by* Wallace and Huffman (2012), Tim Stallmann (2013) boiled the crux of how currently technology-driven electronic maps are frequently missing design, down to the essence, thus corroborating Ken Field's statements about "*cartomyopic musings*" (Field 2013). Basing his statements on the model of *trichotomy of the cartographic process*: 'Build right thing', 'Build thing right' and 'Build thing fast', the Editor of The Cartographic Journal railed against the diktat of "ever-faster" in cartography and struke a blow for artistic, aesthetic attitudes in map making, thus claiming that still 'a few decades back ... many map-makers were artists first and foremost.' (Field 2013: 3).

> "Cartography," wrote Denis Wood in 2004 "is dead." He argued that cartography is increasingly irrelevant as a profession, as a mode of claiming power over the land and as a set of stylistic devices—the everpresent North arrow and scale bar, the neat line boxing off map from page. In 2012, it isn't difficult to find support for his claims. Since the introduction of computer-based geographic information systems in the 1960s, paper maps have mostly ceased being the authoritative storage medium for geographic information. Their role in military conquest and political control has been supplanted by, among other things, the triangulated irregular network elevation data, used in geographic information system (GIS) software, which help guide drone missile strikes. The navigational maps that many people in the United States rely on presently (on smartphones, on the Web and in global positioning system devices) lack neat lines and North arrows completely, and they flout the stylistic conventions of traditional cartography. [...] The North American Cartographic Information Society's inaugural *Atlas of Design* lavishly demonstrates the ongoing vitality of mapmaking, if not the profession of cartography (original emphasis; Stallmann 2013).

Chapter 5
Critical Cartography in the Context of Post-Modernism

5.1 Introduction

This chapter analyses cartography and mapping in the so-called postmodernistic period that challenges the positivism-empiricism and logical positivism approaches of modernity. The emergence of the critical cartography movement from social critical theory since the 1980s is presented in this chapter. Jeremy Crampton, John Krygier, and Chris Perkins analysed the new practices in cartography in the context of information and communication technology (ICT). In the field of critical studies, the contribution of J.B. Harley about cartography as social construction is considered in detail. He draws upon Jacques Derrida's and Michel Foucault's statements about deconstruction and power-knowledge, respectively. This has important implications in cartography, especially for the analysis of maps as 'cartographic text' and as devices or artefacts of power. In summary, the aim of the chapter is to show the relationships between cartography and social theory, and the new viewpoints which can be considered as a paradigmatic shift in the Kuhnian sense in comparison with the traditional or 'more scientific' approach.

5.2 Humanities and Cartography

After the turn of the millennium the export process of cartography from its primary application fields to forms and models of acquisition, to representation and comprehension, which are usable in various cognitive situations and directions of humanities/cultural sciences, have been experiencing an affirmation in philosophy, social sciences, history of sciences, aesthetics, and theory of literature (Besse 2013). Independent of its contents—and apart of its pictorial nature—a map (or atlas) represents an extremely effective form of *writing* (cf. Jameson 1990, Daston and Galison 2007). As a graphic and editorial form of visualisation, organisation, knowledge generation, images and objects, but also as a place of their conservation, an *atlas* has to be considered and used as a *material* as well as a *mental*

P. I. Azócar Fernández and M. F. Buchroithner, *Paradigms in Cartography*,
DOI: 10.1007/978-3-642-38893-4_5,

dispositive determined to facilitate and organise the *legibility* of the world (Besse 2013).

In his paper from 2010 the French philosopher and historian Jean-Marc Besse rises the question of the potential of a *cartographic problematisation* of cultural phenomena, of artistic activities, philosophical reflections and science history (Besse 2010). What, asks Besse, would be the status of cartography, if it would be applied to the fields of art and science history? He believes that there cartography is not only needed with respect to '*territorial relation*' (Besse 2013). In a new orientation of historical research, cartography has to be counted to the *graphic instruments* of various types that permit human spirit and society in general to classify and represent information, data, new findings and ideas in spatial form. In the same way, cartography is a tool for analysis, a classification system and a collection of the cognitive practices which, among others, operate using the means of localisation and spatialisation (Besse 2013: 111). According to Rogoff (2000) cartography is an *epistemic category*, a form of thinking and of historical discovery.

Beside John Brian Harley and Christian Jacob also Jean-Marc Besse states that maps, like other *figurative methods*, are indispensable tools to '*shape*' the world knowledge or also the geographers' reflections about the world (Harley 2001; Jacob 2005; Besse 2008).

5.3 Towards a Postmodern Cartography

Several authors have defined *postmodern cartography* as the period of the discipline that began in the 1980s (Crampton 2001; Crampton and Krygier 2006; Kitchin et al. 2009). Since then, cartography and mapping have been seen from different points of view and perspectives, which differ significantly with respect to the period of modern cartography or traditional cartography.

The key feature of all postmodern thought is a distinct opposition to, or break from, modernism (Ortega Valcárcel 2000). According to Ortega Valcárcel, postmodern criticism is based on the critique of modernity. Here, proposals, worldviews, theories, rational, and scientific grounds are criticised. In other words, the *universalistic discourse*[1] of modernity is criticised, and cartography and mapping are not immune to this new trend.

Nikolas Huffman (1996) discussed the differentiation between postmodernism and the modern period or modernism, dividing the many uses of these terms into four categories[2]: postmodern style in architecture, art and literature; postmodern social theory; the political economic of late capitalism and; poststructural philosophy. He analysed each of these categories and their impact on cartography and

---

[1] In philosophical terms, a universalistic discourse contends that there is a unique reality which can be made known to us through of science, especially factual sciences.

[2] For more details about tendencies during postmodern period see Huffman (1996).

mapping. However, the two categories more important for the development of cartography during the new times are: postmodern social theory and poststructural philosophy.

The postmodern architectural style, emerging in the late 1960s in the wake of the international style of modernism, is also important. So, Huffman (1996) relates it to new styles of cartography by asking: what would a stylistically postmodern map look like? In doing so, he points out that

> The ultimate goal is not to confuse or disorient to the readers, but to encourage them to read deeper into the map and the mapping process, and to challenge the objective and scientific mystique of the *map as mirror of the world* (emphasis added, Huffman 1996: 38).

This statement by Huffman[3] shows the philosophical nature of this postmodern period with respect to modernism. A new perspective arose which makes a powerful criticism to the scientific, cultural, and philosophical bases of modernism. Thus, the metaphor of the map as a reflection or mirror of reality is questioned.

Nevertheless, there are some authors who do not have a radical position in this modernism-postmodernism debate. From the viewpoint of human geography, Laurence Berg, for instance, says that the hegemonic representation of the debate between modernism and postmodernism has contributed to the "two perspectives as unitary, monolithic and opposed essences" (Berg 1993: 503). However, he suggests that an intermediate position between modern and postmodernistic approaches should be found. He recognizes that certain postmodern and some modern discourses are closely related and that these may not differ so radically. Consequently, modernism and postmodernism coexist next to each other in a continuum, rather than in a binary logic and in false dichotomies.

5.4 Critical Cartography

Since the 1990s the field of cartography has been flourishing with writings that identify maps as *social* issues and expressions of *power and knowledge* (Edney 2007). The precursor of this trend was the Englishman (John) Brian Harley (1932–1991), a theoretician in the history of cartography, who considered cartography not only as a science that describes places by means of maps, but also as a science that is capable to contributing a social dimension (Cosgrove 2007). Some of his eminent writings were compiled in the book "The New Nature of Maps: Essays in the History of Cartography" in 2001. After Harley's death in 1991, leading figures that have picked up from where he left off include Denis Cosgrave, Denis Wood, Jeremy Crampton, John Krygier, Marianna Pavlovskaya, and John Pickles (Perkins 2003).

---

[3] Here, Huffman is drawing on the work of Wood (1992).

The aim of critical cartography is to reduce the gap between a technically oriented map design and the theoretical analysis of power in society. In this sense, critical cartographers do not aim to invalidate maps, instead, the criticism is a careful analysis and identification of map *attributes* that are usually taken for granted (Crampton and Krygier 2006). Critical cartography often invokes social theory in order to examine categories of knowledge such as race, territory, boundaries, or identity that are produced or reproduced by maps. For this reason, critical social theory seeks to *problematise mapping* as the social practice linking relationships between mapping and the exercise of power (Harley 2001).[4]

Often the 'production of space' is mentioned (e.g. Casti 2005). In this sense, critical cartographers argue that 'mapping creates a specific spatial knowledge and meaning by identifying, naming, categorizing, excluding, and ordering' (Crampton 2010). Crampton made a distinction in which

> Critical cartographers do not argue that physical space is produced by the process of mapping, but rather that new ways of thinking about and treating space are produced (Crampton 2010: 46).

In addition, Crampton and Krygier (2006) define critical cartography as a 'one-two punch' of new mapping practices and theoretical criticism. For Crampton and Krygier, critical cartography challenges academic cartography by linking geographic knowledge with political power—and is therefore political. They also say that criticism is not a project of finding faults and errors, but an examination of the assumptions in a field of knowledge. Thus its purpose is to understand and suggest alternatives to the categories of knowledge that are used. So, criticism does not seek to escape from categories, but rather to show how they came into being and what other possibilities there are.

In their 2006 article, Crampton and Krygier answer the question 'what is critique?' They refer to the critical theory of the Frankfurt School.[5] They point out that for Michel Foucault,

> Knowledge was established and enabled through historically specific power relation' [However] 'Foucault's conception of power was more subtle, one that emphasized the politics of knowledge. Power did not emanate from the top of a class hierarchy, but rather was diffused horizontally in a highly differentiated and fragmented fashion (Crampton and Krygier 2006: 14).

---

[4] These relationships between cartography and power are exemplified by Harley in his articles such as "Maps, Knowledge, and Power" (1988), "Power and Legitimation in the English Geographical Atlases of the Eighteenth Century" (1997) and "New England Cartography and the Native Americans" (1994) (cf. Harley 2001).

[5] The Frankfurt School was founded in Germany in 1923 and moved to New York in 1933 when Adolf Hitler came to power. Its main exponents are Max Horkheimer, Theodor Adorno, Walter Benjamin, Herbert Marcuse and Jürgen Habermas. They sought to release the emancipatory potential of a society repressed by technology, positivism, and ideology to dispel harmful and illusory ideologies by providing an emancipation philosophy which could challenge existing power structures (Crampton and Krygier 2006).

The authors state that the term *critique* stand for a politics of knowledge:

> First, it examines the grounds of our decision-making knowledge; second, it analyses the relationships between power and knowledge from a historical perspective; and third, it resists challenges and sometimes overthrows our categories of thought. … Furthermore, the purpose of critique as a politics of knowledge is not to say that our knowledge is not *true*, but that the truth of knowledge is established under conditions that have a lot to do with *power* (original emphases by Crampton and Krygier 2006: 14).

Crampton and Krygier also identify two areas in critical cartography where the traditional disciplinary modes of cartography have come in question: on one hand, a theoretical enquiry which seeks to examine the social relevance of mapping, its ethics, and power relations; on the other hand, the development of open-source and pervasive mapping capabilities, in which new kind of users critique the relevance of cartography's of the end of twentieth century.

Jeremy Crampton (2010) named these movements: theoretical critiques and critical mapping practices. In the former, the critique comes from *inside* cartography, but the critics draw subject from other disciplines. In the latter, however, the focus is almost entirely from *outside* the field of academic cartography.

This movement is the "one-two punch" called *undisciplined cartography* by Crampton and Krygier (2006). That means that these two trends, theoretical critiques and critical mapping practices respectively, resist and challenge the received method and practice of mapping that had been established when cartography became an academic discipline. For this reason, and according to the above authors, cartography is "undisciplined": freed from the confines of the academic and opened up to the people.

The first trend developed by critical authors in the context of cartography as social construction will be analysed later in this section. The second trend, which is discussed below, corresponds to the so called 'new practices' in cartography.

5.4.1 Critical Cartography and its New Practices

Before indicating the new cartographic practices, it is important to mention that Denis Wood and John Krygier (2009) have presented the topic of critical cartography in a historical perspective *within* cartography. According to them, academic geographers and cartographers believe that critical cartography is a recent academic phenomenon developed during the late 1980s and early 1990s. Criticism, however, more broadly defined and understood, has been part of mapmaking from its earliest days. In other words, critique has always existed in the history of cartography.

For instance, the authors mention Mercator's critique of both the Ptolemaic Conical Projection, popularised during the Renaissance, and the plane charts, (known as *portolanos*), long used by sailors. Later, the Mercator projection was criticised by Johann Heinrich Lambert (in 1772), James Gall (in 1855), and recently by Arno Peters (in 1974). These were emphatic critiques embedded in novel ways of making maps, in novel map subjects, or both.

Wood and Krygier further point out that the profession of cartography has also been criticised from within. They refer to the internal critique mentioned in Max Eckert's volume *Die Kartenwissenschaft* (Map Science, 1921–25), in Arthur Robinson's textbook *Element of Cartography* (1953), and in Erik Arnberger's *Handbuch der Thematischen Kartographie* (Manual of Thematic Cartography, 1966). All these critiques were directed towards transforming cartography into a science. Some of them were based on the methods that were used (mainly with regards to the psychological empirical tests).

Wood and Krygier address the comments from Wood, Fels, Harley, Woodward, Rundstrom, and Pickles, stating:

> Overturning the paradigm of Eckert and Robinson by shifting attention from the *form* of the map, with which the profession was obsessed, to its *meaning for behaviour*. Instead of asking whether the brain was overcharged by the density of symbols, these critics asked how the body of subject was constructed by the map, that is, how the map oppressed, subjugated, or otherwise impinged on people (original emphasis by Wood and Krygier 2009: 6).

In a similar way, Perkins (2008) argues that in critical cartography, map use is best interpreted by applying methodologies from the social sciences, employing a mixture of ethnographic and textual methods.

Beside the internal critique, there is an external one. Wood and Krygier (2009) give several examples of indigenous communities (especially in North America), who build their own maps with the help of new geo-technologies. *First Nations* or *Indigenous Mapping* offers a critique of the official mapmaking with respect to its prerogatives, its form, and its content. First Nations Mapping has also ties to other forms known as *counter-mapping*. Counter, means: the mapping of professional cartographers. Wood and Krygier mention, among others, new projects, such as ethno-cartography, eco-mapping, bioregional mapping, community mapping, public participation GIS (PPGIS), participatory rural appraisal, green mapping, and Parish mapping.

All these new *performances* have been used by artists.

> Map artists do not reject maps. They reject the authority claimed by professional cartography uniquely to portray reality as it is. In place of such professional values as accuracy and precision, art maps assert values of imagination, social justice, dreams, and myths; … Artists insist that their maps chart social and cultural worlds every bit as *real* as those mapped by professional cartographers' (original emphasis by Wood and Krygier 2009: 9).

The above statement is another example of the undisciplined cartography mentioned by Crampton and Krygier (2006). In this context, art maps contest not only the authority of professional mapmaking institutions (government, business, academia, science), but they also reject the world that such institutions bring into being. Thus, the project of art mapping is nothing less than the remarking of the world (Perkins 2003).

In the same sense, Kitchin et al. (2009), while defining that critical cartography is avowedly political in its analysis of mapping praxis, also assert that critical cartography is,

> however, decidedly not against maps, but rather seeks to appreciate the diverse ways in which maps are produced and used by different individuals and groups (Kitchin et al. 2009: 12).

As example of individuals and groups producing and using maps, Olga Paraskevopoulu, Dimitris Charitos, and Charalampos Rizopoulos (whose background is from outside of the field of cartography), indicate alternative ways of mapping, especially in urban landscapes. These alternative mapping practices also challenge traditional cartography, where new location-sensing technologies such as Global Position System (GPS) are being employed in these progressive practices (Paraskevopoulou et al. 2008).

The authors mention several projects that question and criticise location detection technologies combined with traditional mapping techniques, such as *Urban Tapestries* (in 2002), *Bio Mapping* (in 2004), and *Amsterdam Real Time* (in 2002). These projects intend to raise public concern about the accuracy and ethics of these technological applications (see Paraskevopoulou et al. 2008). They also categorise location-specific art projects with respect to two mapping technologies: namely *spatial annotation* and *tracing mapping*.

For Paraskevopoulou and co-authors,

> All these projects employ location-aware technologies in an attempt to re-attach aspects of the everyday life to urban space either by embedding information and/or emotions on this space or by using the trails of humans or objects for representing spatial events (Paraskevopoulou et al. 2008: 7).

In other words, new aspects of reality are mapped. These new 'objects' have traditionally not been considered by professional or scholarly cartography, and therefore, these alternative cartographies can be considered new practices.

In this respect, Chris Perkins has reviewed collaborative community mapping in the United Kingdom. He details local alternative cartographies, such as Parish mapping, Green maps, artistic map, Open source mapping, and cycle mapping (Perkins 2007). All these alternatives can be defined as *local mapping* that is produced collaboratively by *local people*. Mapping practices mostly employ geospatial technologies (GPS, GIS, digital cartography). With the support of these tools, community mapping offers new possibilities of emancipation activities[6] for marginalised groups.

Similarly, Sophia Liu and Leysia Palen analysed the rise of *map mashups*[7] in the context of crisis information management. They have conducted a qualitative study that considered several of the crisis map mashups for mapping hazards and disasters

---

[6] In the context of geospatial information, emancipation activities refer to all those practices that certain social groups utilize for major access and use of spatial information supported by new information technologies. Previously, this information had a restricted use and was administered only by the government and state organisations with an official character.

[7] A mashup is a website that combines two or more sources of content into one tailor-made experience. Then, map mashups combine or'mash up' multiple sources of data which are displayed in some geographical form. All this is done using application programming interfaces (APIs) and extensible markup language (XML) in the context of Web 2.0 technology (Liu and Palen 2010; Haklay et al. 2008).

Table 5.1 New cartographic practices challenging professional and scientific cartography. For literature references and details see text of Sect. 5.3

New practices	Cartographic projects (place and year of initiation)	Type of maps
Counter-mapping	Parish Mapping (England, 1996)	Map mashups (e.g. Crisis map mashups)[a]
Ethno-cartography	Green Maps (New York, 1992)	
Community mapping	OpenStreetMap (United Kingdom, 2004)	
Public Participation GIS	Cycling Maps (United Kingdom, 2005)	
Participatory mapping	Urban Tapestries (London, 2002)	
Locative media	The PDPal (New York, 2002)	
Participatory 3-D modelling	Bio Mapping (London, 2004)	
Collaborative mapping	Amsterdam Real Time (Amsterdam, 2002)	
	Cabspotting (San Francisco, 2006)	
	MILK (Netherlands, 2003)	

[a] Examples of "Crisis Map Mashups" were compiled by Liu and Palen (2010)

(Liu and Palen 2010). They claim that "crisis map mashups" are becoming a *neo-cartography* because of these new tools that have emerged among nonprofessional cartographers in the context of the emergence of neo-geographic practices.

In summary, the following table shows some of the new cartographic practices that have arisen during the critical cartography period. These practices challenge the conventions and rules of the modern cartographic scientific approach. Sometimes the terms community mapping, participatory mapping, and collaborative mapping are considered synonymous. In Table 5.1 some cartographic projects (cited by Perkins 2007; Paraskevopoulou et al. 2008), and the main types of maps are shown. These maps have been created with the participation of local people and local communities using geospatial technology and Geoweb.[8] As can be seen in this table most of the new cartographic practices expanded during the decade of the 2000s, although the first project had already emerged at the beginning of the 1990s.

In conclusion, this assault on the presumptions of professional cartography—or modern cartography—extends to its most fundamental categories of knowledge, according to the examples mentioned above. For that reason this movement called critical cartography together with the new cartographic practices points us toward epistemological and philosophical implications for the discipline.

5.5 Cartography as Social Construction

Some studies in cartography continued to implicitly adopt *scientific approaches* concerned with improving the efficiency of the way in which maps communicate: in other words, Robinson's legacy (Perkins 2003). This implies the comparison of

[8] The term GeoWeb or Geospatial Web implies the merging of geographic (location-based) information with the abstract information that currently dominates the Internet (Haklay et al. 2008). The emergence of the Geospatial Web, particularly Web Mapping 2.0, had led to increases in geobrowsing activities (e.g. browsing through Google Maps or Google Earth).

different designs of the same map, comparative analyses of map designs in different genres, and different user perceptions of mapping products. Nevertheless,

> These empirical studies almost all implicitly take an a-theoretical view of the map as a "mirror of the world", or as part of a system of cartographic communication, largely isolated from social and historical context (Perkins 2003: 334).

Additionally, Della Dora (2009) in her article about performative atlases,[9] states that:

> Matthew Edney (1993), Christian Jacob (2005), and Jeremy Crampton (2001) have argued for a history of cartography "without progress", in which maps are no longer assessed in terms of accuracy versus inaccuracy but, rather, are valued as "social constructions" that are always contingent on the specific cultural, social, and technical relations at different places and times (Della Dora 2009: 241).

From the above statements, a clear difference from the traditional or modern cartography can be seen. The view that cartography produces objective, neutral, and scientific maps has been challenged. At the end of twentieth century there is an alternative cartography in which maps are recognized as social constructions. So the map is no longer the mirror of the reality from an objective viewpoint.

Similarly, Crampton pointed out that recent developments in cartography have gone well beyond the model of maps as communication. He describes these developments

> as an *epistemic break* between a model of cartography as a communication system, and one is which it is seen in a field of power relations, between maps as presentation of stable, known information, and exploratory mapping environments in which knowledge is constructed (original emphasis by Crampton 2001: 253).

An important contribution made by the postmodernist approach is that a map can be seen as a *text*; hence, it is not an objective form of knowledge, but rather has numerous hidden agendas or meanings. Brian Harley, a historian of cartography and maps, elaborated this perspective in cartography on topics such as 'maps, knowledge and power', 'deconstructing the map', and 'silences and secrecies in maps'.

Harley's main papers were published in a book (Harley 2001), and give the vision of the map as *a social construction*. He suggests a new research agenda for cartography by arguing that maps play an important role in different societies and that they often reinforce the *status quo* or the interest of power. He also suggests (especially in the history of cartography) to research the historical and social context in which maps are created and employed.

---

[9] Atlases are conceptualised as mnemonic tools activated through different types of personal encounters (always contingent) that are at once visual and tactile. "As a mnemonic device, the atlas is thus always *in process* through a diverse set of interrelated practices .... Exploring the atlas means transforming these micro-performances into a mental movement of the entire body over territory" (Della Dora 2009: 249).

Harley proposes an epistemological shift in the way of interpreting the *nature of cartography*. He questions the premise that cartographers are grounded in scientific or objective knowledge creation. Thus, the question arises whether the concept of a *progressive science* is a myth created largely by cartographers in the course of their professional development through the normative models of cartography (Harley 2001).

The belief in progress in cartography means that through the application of a scientific perspective we can make representations of reality which become more and more objective and accurate (Edney 1993).

For Harley, the link between reality and representation has dominated cartographic thinking. This linkage has also led cartography away from the path of normal science as it has been practiced since the Enlightenment. This relation has offered a ready-made epistemology and has been 'taken as fact' in the history of cartography (Harley 2001). These ideas about objectivity in cartography and mapping also have been debated in 1996 by Matthew Edney in his 'Theory and the History of Cartography'. 'Cartographic history has been dominated by an empiricism that treats the nature of maps as self-evident and which denies the presence of any theory' (Edney 1993). The aim is to suggest an alternative epistemology, embedded in social theory rather than in scientific positivism, which is more adequate for the history of cartography. In this context, Harley proposed the *deconstructionist* method (described in the next section) to break the association between reality and representation.

5.5.1 Deconstruction and Cartographic Text

Deconstruction is an approach, which rigorously pursues the meaning of a text to the point of undoing the oppositions on which it is apparently founded, and the point of showing that those foundations are irreducibly complex, unstable, or impossible. Deconstruction has had an enormous influence in psychology, literary theory, cultural studies, linguistics, feminism, sociology, and anthropology (Reynolds 2002). Deconstruction, in simple terms, is a notable postmodern technique that enables the researchers to uncover hidden meanings and agendas (Hallisey 2005).

In the context of cartography, Brian Harley developed three threads of argument in the pursuit of the deconstructionist strategy (Harley 1989). First, he examines the *cartographic discourse* under the light of some of Foucault's ideas[10] about the play of rules within discursive formation. Then, he draws one of

[10] Michel Foucault played a critical role in the development of the postmodern perspective that knowledge is constructed in concrete historical situations in the form of discourse; knowledge is not communicated by discourse, but is discourse itself, and can only be encountered textually. Foucault performs what he calls "genealogies," attempts at deconstructing the unacknowledged operation of power and knowledge, to reveal the ideologies that make domination of one group by another seem "natural" (Brewton 2002). For more information on the Foucauldian discourse applied in geography, cartography, and mapping see Crampton and Elden (2007).

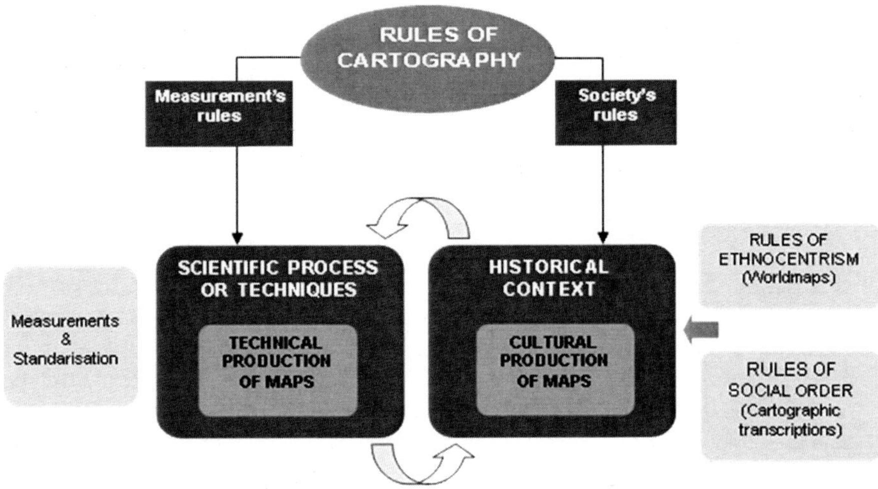

Fig. 5.1 Rules of cartography according to Harley (1989), considering the Foucauldian discourse approach

Derrida's central positions to examine the *textuality of maps* and its rhetorical dimension. Finally, he returns to Foucault to consider how maps work in society as a form of *power-knowledge*.

According to Harley (1989), one of Foucault's primary units of analysis is discourse, which has been defined as a system of possibility for knowledge (Harley 1989: 3). So, Harley related *discourse* with *rules,* and he asked a question about which type of rules has governed the development of cartography (see Fig. 5.1).

Brian Harley pointed out two distinctive sets of rules that underlay and dominate the history of Western cartography since the seventeenth century. One set may be defined as governing the technical production of maps and the other set relates to the cultural production of maps (see Fig. 5.1).

So, the first set of cartographic rules can be defined in terms of scientific epistemology (e.g. scientific rules, measurement rules). For the second set of rules, Harley gives two instances: "rule of ethno-centricity" and "rules of social order" or rules of hierarchy of the space or territory (for more details see Harley 1989). He claims that

> To discover these rules, we have to read between the lines of the technical procedures or of the map's topographic content. They are related to values, such as those of ethnicity, politics, religion, or social class, and they are also embedded in the map-producing society at large (Harley 1989: 5).

The author suggests that these rules operate both within and beyond the orderly structures of classification and measurement, and they go beyond the stated purposes of cartography. Harley notes that the interplay between social and technical rules is a universal feature of cartographic knowledge:

> In maps it produces the "order" of its features and the "hierarchies of its practice". In Foucault's sense, the rules may enable us to define an *episteme* and to trace an archaeology of that knowledge through time (original emphasis by Harley 1989: 6).

Concerning the rhetorical and textual dimensions in the context of cartography as social construction, Harley draws from Jacques Derrida that the model of text can be applied to other types of texts and not necessarily only to literary texts. For Harley, maps communicate as much as to provide a powerful rhetoric, and therefore can be critically examined as texts themselves. So for him, 'text' is a better metaphor for maps than 'the mirror of nature' is. Maps are *cultural texts*. By accepting their *textuality*, we are able to embrace a number of different alternative possibilities in reading and meaning (Harley 1989).

Deconstruction, as discourse analysis in general, demands a closer and deeper reading of the cartographic text than has been the general practice in either cartography or the history of cartography. For Harley, the philosophy of deconstruction by Derrida[11] to broaden such interpretation to all maps is helpful.

According to Brian Harley we can show how the cartographic (i.e. scientific) facts act also as a symbol (i.e. a metaphor). This means that in "plain" scientific maps, science itself becomes the metaphor. He mentions several examples in which we can trace the contours of a metaphor in a scientific map, and understand how the text works as an instrument operating on social reality (Harley 1989).

Furthermore, the play of rhetoric in the deconstructionist theory is close to that of the metaphor. Harley argues that, notwithstanding the efforts of scientific cartography to convert culture into nature and to naturalise social reality, it has remained an inherently rhetorical discourse. Therefore, for him "there is nothing revolutionary in the idea that cartography is an art of persuasive communication (Harley 1989: 11)".

Harley's position is to accept that rhetoric is part of the way in which all texts work and that all maps are rhetorical texts. In turn, this stance is carried out to cartographic terms; for example,

> The steps in making a map (selection, omission, simplification, classification, the creation of hierarchies, and "symbolization") are all inherently rhetorical. In their intentions as much as in their applications they signify subjective human purposes rather than reciprocating the workings of some *fundamental law of cartographic generalization* … The issue in contention is not whether some maps are rhetorical, or whether other maps are partly rhetorical, but the extent to which rhetoric is a universal aspect of all cartographic text (original emphasis by Harley 1989: 11).

We claim that these statements exemplify maps as a social construction and also challenge the scientific stance of cartography which was maintained until the

---

[11] In his analysis Harley cites Jacques Derrida's book *On Grammatology* (translated in 1976). In this book Derrida introduces the term *deconstruction* to describe the manner that understanding language as writing (in general) renders infeasible a straightforward semantic theory (Derrida 1976).

late 1980s: maps as a mirror of the world. In Kuhnian terms, it is a new viewpoint or another worldview. This change in the map conception and its interpretation can be considered as a paradigmatic shift from within the discipline.

5.5.2 Power and Maps

Throughout the history of cartography, maps have been understood as an *instrument of power*[12] for states and municipal administrations, strategic military planning units, colonialism, and any centralised power structures. Maps are powerful tools. They are created by those who have authority and power, and for this reason they are powerful. Thus, maps are important documents or tools about the possession of a territory or an area, or the knowledge about the location of certain places of value.

Jeremy Crampton (2010) points out that mapping is embedded in a specific relation of power:

That is, mapping is involved in *what* we choose to represent, *how* we choose to represent objects such as people and things, and *what* decisions are made with those representations (original emphases by Crampton 2010: 41).

Chris Perkins mentions different authors who have investigated the power-relation in mapping. He also notes different areas in which maps act as power, such as mapping in imperial projects, in commercial *élites*, military power and geo-politics, and in property relationships (Perkins 2003).

Brian Harley has to be considered one of the pioneers of the history of cartography who showed the power relation in maps, and consequently the power of cartography as a social practice. He drew concepts such as power-knowledge from social sciences to cartography (cf. Harley 2001).

In the context of how maps work in society *as a form of power-knowledge,* Harley (1989) proposed a distinction between *external* and *internal* power in cartography. It helps to understand how power works through cartographic discourse and the effects of that power in society (Fig. 5.2).

On the one hand, the most familiar sense of power in cartography is that of the *external power* to and of maps and mapping. Here power is exerted *on* cartography and power is also executed *with* cartography. It is an external power, often centralised and exercised bureaucratically, imposed from above, and it manifests itself in particular acts or phases of deliberate policy (especially centres of political power: monarchs, ministers, and the states; Harley 1989).

On the other hand, Harley defines the *internal power* of cartography very differently. Here, the focus of inquiry shifts from the place of cartography in a

[12] The relationships between maps and power are not unique to postmodern cartography. Throughout history, the practical use of maps as means of power has been documented; however, the novelty is that this relation began to be systematically investigated from the second half of the 1990 onwards. For this reason, the section "Power and Maps" has been included within the cartography of the postmodern period.

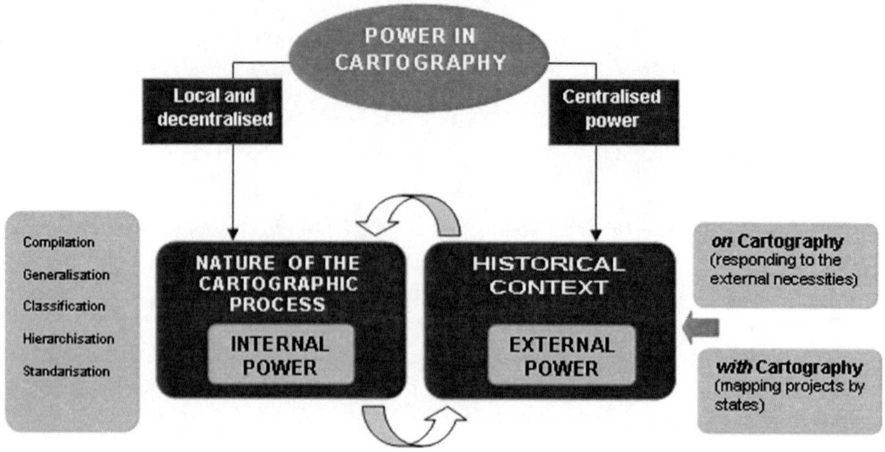

Fig. 5.2 Power of cartography according to Harley (1989), considering a Foucauldian power-knowledge approach

juridical system of power, to the political effects of what cartographers do when they make maps.

> Cartographers manufacture power... It is a power embedded in the map text: we can talk about the power of the map just as we already talk about the power of the word or about the book as a force for change. In this sense maps have politics (Harley 1989: 13).

Similarly, power comes from the map and it traverses the way maps are made. The key to this internal power is the *cartographic process*. So, to catalogue the world is to appropriate it; all technical processes represent acts of control over its image which extend beyond the professed uses of cartography. Thus, the world is disciplined and normalised.

Harley exemplifies the *standardisation* and *normalisation* of the world through an analogy between factories and cartographic workshops. In the former, the goods manufactured are standardised and in the latter, *world's images* are standardised by cartographers. There exists also another analogy between scientific laboratories and maps: In the first, the explanation of the processes of the physical world is created with formulas, and in the second, nature is reduced to a *graphical formula*.

Taking into account Foucault's considerations, Harley does not suggest that power is deliberately or centrally exercised. He says that it is a *local knowledge* which at the same time is *universal* and usually passes *unnoticed*. To consider the effects of cartography in society, or the 'logic of the map' upon human consciousness, Harley suggests that

> We have to consider for maps the effects of abstraction, uniformity, repeatability, and visuality in shaping mental structure, and in imparting a sense of the places of the world (Harley 1989: 13).

From an epistemological viewpoint, Harley pointed out that whilst the map *is* never the reality, it helps *to create* a different reality. Once embedded in the published text, the lines on the map acquire *authority*. Thus, maps are authoritarian images, and even without the users being aware of it, a map can reinforce and legitimate the *status quo* inside of a society.

Figure 5.3 summarises Foucault's and Derrida's contributions that were incorporated into cartography by Brian Harley. It shows by means of the external and internal power of cartography and the types of rules that have governed cartography through time, that the *cartographic discourse* acquires a strong power.

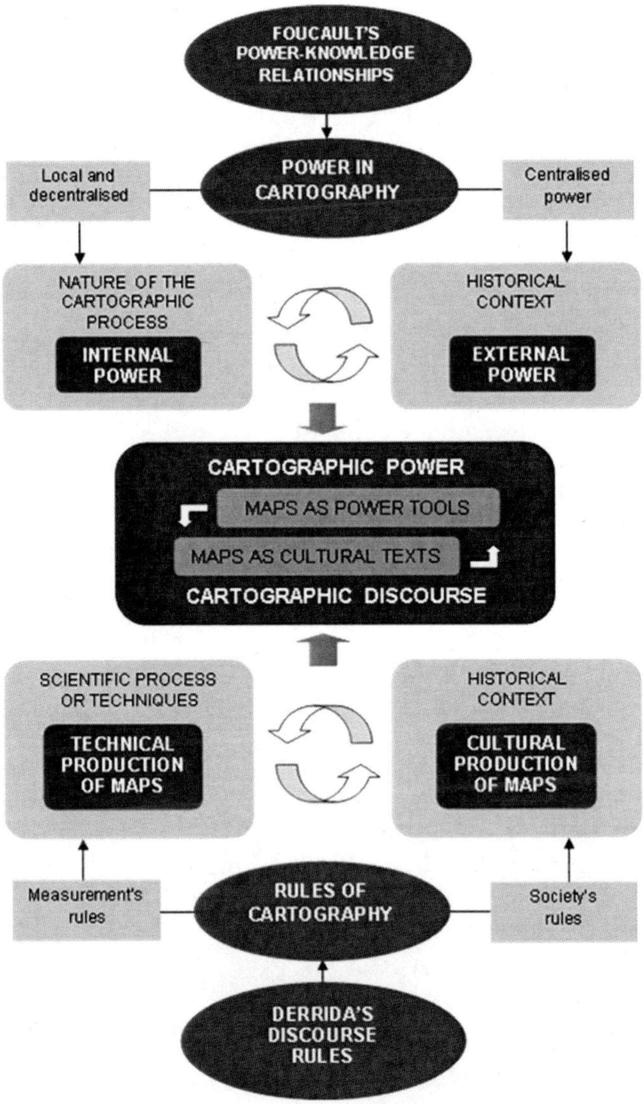

Fig. 5.3 Power play of the cartographic discourse according to Harley (1989)

Therefore the cartographic power and the cartographic discourse are mutually reinforced through *maps as tools of power* and *maps as cultural texts,* respectively.

This approach was originally applied to old maps (Harley 1988a, 1988b, 1994, 1997), but the social implications of maps have been a major topic in analysing modern maps. Authors such as Denis Wood and Mark Monmonier have worked on these topics, especially in their books *The power of maps* (Wood 1992) and *How to lie with maps* (Monmonier 1996).

Similarly, Gyula Pápay (2006) points out that on the political level regarding the relationship between politics and cartography, the emergence of a complete sense of possession has had an ambivalent effect in history as a result of maps. He writes that

> On the one hand, maps were required as instruments of the preservation of possession or power; on the other hand, they were kept secret to ward off foreign claims (Pápay 2006: 1).

In this sense, maps play a double game: they can document claims to power and are thus a means of the presentation of power, and they can also be used to deny power by means of hiding of cartographic documents. For example, Pápay (2006) examined the historical relationship among politics, secrecy, and the political maps.

5.6 Neo-Cartography: A Concretion Towards the Map as Social Construction

Maps had generated power throughout history. At the beginning of the twentyfirst century this classic power is about to be transferred to the regional or local communities. As mentioned in the previous section, this *map power* was in the hands of formal and traditional institutions which managed official cartography and mapping. Making us of the upcoming new technologies arose an agenda for democratising cartographic visualisation at the end of twenty-century (Rod et al. 2001). This way, in the context of advanced computer technology, a second generation of users developed who are not aware of traditional cartographic rules or cartographic standards.

At the 25th International Cartographic Conference in Paris in 2011, the Commission on Neocartography was accepted as a commission of the International Cartographic Association ICA. This commission's terms of reference state that

> many examples of new and innovative mapping are being produced outside the normal orbit of existing cartographers or map producers. The term neocartography is being used to describe map makers who may not have come from traditional mapping backgrounds, and are frequently using open data and open source mapping tools. Another difference is in the blurring of boundaries between map producers and maps consumers. The availability of data and tools allows neocartographers to make their own maps, show what they want, and often be the intended audience as well – that is to say they may make the maps for themselves, just because they can. There is a real need for a discipline to be established to

> study this essentially undisciplined field of neocartography. (Commission on Neocartography 2011, digital text: no pagination).

In a similar way William Cartwright adds that 'the traditional model of formal – mainly governmental – information collection, storage and publishing is now complemented by a less formal and more personal data collection and publishing model'. This type of mapping has been called the 'GeoWeb', 'Volunteered Geographic Information' (VGI)* and 'crowdsourcing' (Cartwright 2010, digital text: no pagination). Also Holger Faby and Andreas Koch mention as examples of the mentioned developments: interactive generation of geo-referenced data and services ('mashups'), collaborative mapping ('crowd sourcing'), real-time information concerning mobile activities ('tracking') and individualised authoring of routes and places ('blogs') (Faby and Koch 2010).

Since its advent, developments related to the Internet had always been drivers for new cartographic applications. 'Maps as interfaces to abundant information systems as well as presentation forms of spatial-related information are offered ubiquitously, either through their application on mobile input and output devices or on the Internet' (Gartner 2010, digital text: no pagination). For Georg Gartner the term 'Neocartography' is used to express the collaborative character of the integrative possibilities in modern Internet Cartography. Associated to the mentioned term there exists also 'volunteered geographic information' (VGI)[13] to express the specific feature of the users' willingness to make information available (cf. Goodchild 2007; Paraskevopoulou et al. 2008; Perkins 2007). Georg Gartner points out that the collaborative and social aspect of the 'new Internet-Cartography' also caused the term "wikification" in connection with mapping (cf. Sui 2007).

This new tendency has also been studied in Latin America (particularly in Brazil) in the context of social cartography: considering relations among geoinformation, citizenship and social participation (Di Maio et al. 2011). The authors claim that

> the right to think about the world and interpret social and territorial local problems may be, by the participatory method and the use of maps and GIS, a force for social and political statement and also an additional ideological mechanism to control hegemonic agents not engaged in the genuine processes of citizen's socialization. (Di Maio et al. 2011: 45).

In the field of application, Mohsen Kalantari and Abbas Rajabifard explored potential uses of crowdsourcing in cadastres. They argument that citizens can contribute towards the identification of inaccurate boundaries especially in developing countries. This crowdsourcing support is another way in which local people can generate cartographic information contributing to social benefit.

---

[13] One might also talk about *volunteered cartographic information* (*VCI*), which, in fact, is a more appropriate term, since we are referring to geo-located graphic, and not only to descriptive or tabular information.

> Public participation not only enables the cadastre to be crowdsourced but also improves its governance; responsive participation of citizens increases a government's accountability, and empowering people enhances poor performance. (Kalantari and Rajabifard 2012: 33).

The ICA Commission on Neocartography claims that

> neocartography should look at the impact of the digital turn on cartographic practice and cartographic principles, just as it should reflect on more theoretical aspects of the ethics and philosophy behind geospatial visualisations (2011, digital text: no pagination).

Along the same line we propose *neo-cartography* as a "paradigmatic shift in its gestation stage". Technological changes can impact the philosophical and epistemological basis of the discipline. However, for the consideration of these changes in Kuhnian terms further aspects have to be analysed. Nevertheless, there is a change in which the mapmakers' "new" generation is contributing to establish maps as social constructs.

5.7 Mapping vs Cartography

In his article of 2005 the American cartographer Denis Cosgrove derives the development of modern *mapping* from the history of cartography, in particular from the differentiation between physical and thematic geography as it developed in Germany during the 19th century[14] (Cosgrove 2005: 28; Dünne 2013: 222). Cosgrove sees a 'much richer and more complex role of mapping/cartography in cultural geography' (Cosgrove 2005: 28) which developed in the decades and in particular in the years before the turn of the millennium. Cosgrove draws the starting point of a new understanding of mapping from Peter Jackson's *Maps of Meaning* (Jackson 1989), a study in which, according to Cosgrove, Jackson makes *the* crucial step by extending the meaning of the term *mapping* metaphorically 'to include all graphic representations of knowledge' (Cosgrove 2005: 28). This, again, serves as a starting point for the *Critical Cartography* of the 1990s from which, among others, arose the still ongoing mega-project of a new comprehensive *History of Cartography* (Harley and Woodward 1987 ff., Dünne 2013). In his preface to Volume 1, the late editor-in-chief John Brian Harley defines maps as 'graphic representations that facilitate a spatial understanding of things, concepts, conditions, processes, or events in the human world' (Harley and Woodward 1987: XVI).

The procedure of *mapping* is, however, not so new. As pointed out by König (2013), already in 1976, hence long before we began to organise our visual communication by means of Windows or Apple platforms, Arthur Robinson and

[14] Cosgrove (2005), however, does not have a more in-depth consideration of the not unproblematic differentiation between a "basic", i.e. physical cartography and a thematic cartography based on the first one Dünne (2013). For more details about the delicate separation of physical and thematic cartography see Stockhammer (2007: 44 ff.).

Barbara Petchenik wrote: 'Mapping is based on systems of assumptions, on logic, on human need, and on human cognitive characteristics' (Robinson and Petchenik 1976: X). Further they write:

> The apparent simplicity of an ordinary sketch map is depictive; in fact, even the simplest map is a remarkably complicated instrument. It is quite reasonable to suppose that the map, as a communicative device, has been around as long as written language has: like writing, a map is a way of graphically expressing mental concepts and images (Robinson and Petchenik 1976: 1).

Thus, like 10 years later the American art historian William J. Thomas Mitchell, define a map as a *medium of communication*: 'A map is a value-laden *image* which is used for communication'[15] (Mitchell 1986: 9–14; emphasis by the authors). The French geographer Christian Jacob even defines maps with more 'easiness' (König 2013):

> The definition What is a map? The question, in all its simplicity, invites for tautologies. [...] The eye does not see, it constructs, it designs the space. The map is not an object but an operation. Mediation, interface, it is a rest of secrecy. (Jacob 1992: 29) [...] Finally, for a long time the map has been a metaphor for the description of human relations, relationships of power, the hierarchical distribution in a social group: an abstract and operational space, reduced to guidelines and motion lines, crossings of borders and passages, and the marking of nodal points (Jacob 1992: 32).[16]
>
> Let us recapitulate: The map does not define itself through what it represents: too diverse are the spaces of cartography in terms of their scale and nature. Neither does not define itself by its fix visual configuration. If a map assigns itself to the category of images, it does not present stable characteristics and is recurring like a landscape portrait or painting. Neither defines the map itself through a unique function. And even less by the geography, the field of knowledge which would justify which depiction serves as preferred means (Jacob 1992: 36–37).[17]

So, Christian Jacob describes *all what a map not is* and offers a rather "open" definition including that of a map as an 'abstract, operative space, thus coming

---

[15] Gyula Pápay also consider maps as images (cf. Pápay 2005a, 2005b, 2005c and Chapter 4.9.3 Cartography and Visualistics).

[16] La definition. Qu'est-ce qu'une carte? La question, dans sa simplicité même, invite la tautologie. [...] L'oeil ne voit pas, il construit, il imagine l'espace. La carte n'est pas un objet, mais une fonction. Médiation, interface, elle est un reste occulté. (Jacob 1992: 29) [...] La carte, enfin, est depuis longtemps une métaphore pour decrier les relations humaines, les rapports de pouvoir, les partages hiérarchiques dans un group social: un space abstrait et opératoire, réduit à des lignes de force et de movement, traverse de frontiers et de passages, jalonné de points nodeaux. (Jacob 1992: 32).

[17] Récapitulons. La carte ne se définit pas par ce qu'elle représente: trop divers sont les espaces cartographies, par leur échelle comme par leur nature. Elle ne se définit pas non plus par une configuration visuelle fixe. Si la carte se situe dans la catégorie de l'image, elle ne présente pas de traits structurels stables et récurrents comme le portrait ou la peinture paysagère. La carte ne se définit pas non plus par une function unique. Et encore moins par la géographie, champ de savoir don't elle ne constituerait que l'illustration fidèle comme l'instrument privilégié. (Jacob 1992: 36–37).

close to today's common application like formulated by Kraak and Ormeling (König 2013: 171):

> The term 'map' is used in many areas of science as a synonym for a model of what it represents, a model which enables one to perceive the structure of the phenomenon represented. Thus mapping is more than just rendering, it is also getting to know the phenomenon that is to be mapped. By 'cartographic method' one understands the method of representing a phenomenon or an area in such a way that its spatial structure will be visualized and this will usually take some experimenting. When representing spatial information in map form one has to limit oneself, on account of the available space, to the essentials, and amongst which is the information's structure (Kraak and Ormeling 1996: 40).

In other words, as concisely phrased as the above two authors did it a few pages further, "Maps are nowadays regarded as a form of scientific visualization" (Kraak and Ormeling 1996: 42).

As already pointed out by Dünne (2013), the particular importance of this extension of the classical map definition is first of all that the *cultural technique* ("*Kulturtechnik*") of mapping (Dünne 2013: 222) is not only alotted to literate cultures, i.e. cultures with an alphabetic writing system, but also to other cultures. This finds its reflection in the non-eurocentric structure of the *History of Cartography*. Also, by detaching the mapping from the history of the mapping operations as a 'mainly colonial-territorial power technique' (Dünne 2013: 222), in this *History of Cartography* cartography is not only understood as a domain of historical sciences or art history but is burgeoned a broader interdisciplinary access to maps. In this context only some of the major studies about the early modern cartography shall be mentioned, such as Conley (1996), Padrón (2004), Stockhammer (2007), Glauser and Kiening (2007) and Dünne (2011).

Jordana Dym and Karl Offen consider 'a map as a graphic presentation of space (*real or imagined*, terrestrial or otherwise) that organizes, presents and communicates spatial information visually' (Dym and Offen 2011: 6; emphases by the authors). In their subsequent definition of maps they go pretty far:

> Maps are simultaneously material and social, real and physical products that reflect the cultural concerns, values, and communication arts and technologies of the society that produced them. Maps also tell stories about people and places they show. Maps literally and figuratively influence the way we see the world [...]. Like all visual arts, maps are communication devices that rely on cultural conventions and assumptions to establish their meaning (Dym and Offen (2011: 6).

As already mentioned by König (2013), Dym and Offen (2011) extensively comment on the still comparatively concise aforementioned definition of maps given by Harley and Woodward in their first volume of *The History of Cartography*. The difference of the two definitions by Harley and Wood and Dym and Offen has consequences for both the understanding and the function of a map: if it 'facilitates' the spatial comprehension (like for Harley and Woodward), then we have to deal with a map in the traditional sense, intentionally produced by the map author(s). If, however, a map "reflects" various contents, facts and contexts in their spatial relations – one may say in their topology – like for Dym and Offen or for ESRI (various promotional publications), then, in the strict sense, every graphic representation or depiction with a spatial relation has to be considered a map (cf. König 2013: 162).

As a consequence, Kraak and Ormeling (1996) differentiate:

> Maps are nowadays regarded as a form of scientific visualization […], and maps indeed existed already before visualization developed into a distinct field. The objective of visualization is to analyse information about relationships graphically, whereas cartography aims at conveying spatial relationships. […] The emphasis in scientific visualization […] is more on its analytical power […] than on its communicative aspects; it is primarily directed at discovery and understanding. In cartography, emphasis can lie equally in analysis and communication (Kraak and Ormeling 1996: 41).

We are, as König quite correctly pointed out, thus operating with two different categories, visualization and mapping of graphically presented relations on one, and cartography on the other side which brings topographic data into relation and allows geographic orientation (König 2013).

5.8 Personal GIS

In a remarkable paper, Amin Abdalla argued that in the second decennium of the 21st century, the time may be mature to personalize not only maps but also GISs in order to enable 'ordinary citizens', i.e. the non-expert users to manage their personal spatial data, to plan activities, and to accomplish certain space-related tasks.

> … many functionalities have already been implemented in various applications; however, mostly independent of each other. For instance have online-maps to be used parallel to calendars to find the places of particular appointments. A [*Personal GIS*] *PGIS*, thus, represents a bundling of spatial and temporal functions in one application plus the creation of new functions through a possibly better representation of information in the form of alternative data structures[18] (Abdalla 2013).

Such a *Personal GIS* may be based on the three columns *support*, *analysis* and *organization*, which may, again, be hierarchically broken down into decision support, execution support, spatial analysis, pattern recognition, and spatio-temporal indexing (Abdalla 2013). It must be expected that the near future will prove that PGIS are adding a new dimension to geoinformation and, hence, to cartography.

[18] Auffallend […] ist der Fakt, dass viel der Funktionalitäten schon in diversen Applikationen umgesetzt wurden; doch sind diese meist unabhängig voneinander. So geschieht es beispielsweise, dass eine Onlinekarte parallel zu einem Kalender benutzt werden muss, um den Standort eines Termins herauszufinden. Ziel eines PGIS ist also die Bündelung von räumlichen und zeitlichen Funktionen in einer Applikation sowie die Einführung neuer Funktionen durch eine möglicherweise bessere Repräsentation der Informationen in Form von alternativen Datenstrukturen. (Abdalla 2013: 88).

5.9 Geocomposition

One particular view about the synoptic use of the available new technologies needs special mention. Around the turn of the millennium, the Polish cartographer Zenon Kozieł began to treat cartographic representation in a very holistic and theoretically well-based way (Kozieł 1997) and subsequently to coin the term *geocomposition* (Kozieł 2001). Under geocomposition Kozieł understands

> a specific semiotic *universum* (semiosis), with a very wide cognitive and utilitarian meaning. It is identified with graphic, sound and text, as well as touch, smell and taste compositions, which relate to the Earth surface and the whole geosphere. The concept of geocomposition covers various elements, devided into: types (anthropogenic i.e. artefacts and natural, so-called geofacts); varieties (real, virtual, augmented); and categories (classical, *curiosum*, metaphoric). The essence of geocomposition as an autonomic semiotic existence is contained in the need to develop interdisciplinary systems of multimedia (ISM), which take into account all the above mentioned elements (Kozieł 2013).

With this definition, the approach of Zenon Kozieł can be brought into the proximity of both Buchroithner's multi-sensory cartographic view of Popper's Three Worlds Model as well as his *VR Cartography* (Buchroithner 1997, 2005; cf. Sect. 2.4.2). In two papers, Kozieł describes his holistic multimedia approach (Kozieł 2005 and 2006). In this context it is interesting that the Russian Alexander Leonov uses the term "composing of geo-images" (Leonov 2001).

Chapter 6
Post-representational Cartography

6.1 Introduction

In this chapter, we try to analyze what the term post-representational cartography implies. Since the first decade of the twentieth century, several authors of social science and social theory have been proposing new map conceptions in comparison with a scientific-empiricism approach. This tendency not only challenges the modern period, but also the critical perspective assumed in cartography and mapping. Here, a group of authors who can be placed in a transitional phase of cartography challenge the traditional proposals: authors such as John Pickles, Martin Dodge, Rob Kitchin, and Chris Perkins propose to analyse cartography and mapping beyond representational knowledge, i.e. accurate and objective maps, and so consider these as a set of spatial practices that do work in the world. Such epistemological foundations, proposals, and challenges to a post-representational cartography are analysed in this chapter.

6.2 What is Post-representational Cartography?

Post-representational cartography is a new perspective in mapping that is contrary to the viewpoint of 'maps as truth' and wants to go beyond the 'maps as social constructions' approach. The former represents the view of modern or traditional cartography, and the latter one is framed in postmodern cartography.

Maps as truth, from a postmodern perspective, implies a cartographic activity that is academic, scientific, objective, and freed of values, and whose aim is to represent reality, i.e. the territory, in an accurate and precise way. In this sense, the metaphor 'the map as mirror of the world' is valid: to depict reality–geographical, spatial, and territorial—and various aspects of it as they are.

In this way, the postmodern approach considers cartography as an academic and scientific pursuit, which has largely consisted of theorising about the best way to represent and communicate that truth. In this quest, the main approach has been the cartographic communication model.

P. I. Azócar Fernández and M. F. Buchroithner, *Paradigms in Cartography*,
DOI: 10.1007/978-3-642-38893-4_6, © Springer-Verlag Berlin Heidelberg 2014

Unlike the scientific approach, Harley (1989) argued that the process of mapping is the *creation of knowledge*, rather than simply revealing it. Therefore, maps are based on the values and judgments of the individuals who create them and so are a reflection of the culture in which they are inserted. In this sense, maps are social constructions, i.e. *expressions of power/knowledge*.

In short, cartographic approaches envision two different ideas about the ontology of maps: maps as objective truths and maps as social constructs. Despite those differences, there are authors who point out that both stances actually conceive maps as *inherent truths*. John Harley, Denis Wood and John Fels note that 'The map itself remains ideologically neutral, with ideology bound to the subject of the map and not the map itself' (cited by Kitchin 2008: 211).

Along these lines Crampton (2001) questions the *ontological grounds* of cartography, identifying the map as a *contingent product* imbued with historical conditions, the time and space of its production and use. Therefore, the map is unable to reflect the truth. Similarly, Kitchin and Dodge (2007) historically reviewed the idea of an ontological crisis in cartography, and they propose going beyond the critique of the ontological basics of cartography. They have called for a radical approach of 'rethinking maps,' as explained in the next section.

6.3 Transition Towards a Post-representational Cartography

In their article "Thinking about maps," Rob Kitchin, Chris Perkins, and Martin Dodge attempt to rethink cartography and mapping in ontological and epistemological way. Considering a post-representational theory of mapping, they summarised the contemporary development of cartography as two stages: from ontic knowledge to ontology, and from ontology to ontogenesis (Kitchin et al. 2009: 10/16).

The social constructivist approach has brought many contributions to the rethinking of maps in cartography. However, some authors claim that the criticism developed by Harley did not go far enough in rethinking the ontological bases for cartography (Kitchin et al. 2009). Authors such as Barbara Belyea (1992), Denis Wood (1993), and Jeremy Crampton (2003) agree that Harley's application of Foucault is limited and incomplete. In general, this new post-representational approach points out that Harley's strategy, i.e. to uncover the ideology inherent in the representation as actual *counter mapping,* does not challenge the ontological status of maps. This simply reveals the politics of mapping: there arises a question about the ideology of the topic of a map, but not about the map itself (Kitchin et al. 2009).

Table 6.1 summarises different map concepts that have arisen during the period of postmodern cartography. It also includes a comparison between the modern cartography approach and the latest period called post-representational cartography. Although it considers the methodology by Kitchin et al. (2009), this table also includes other authors, such as Jeremy Crampton, Emanuela Casti, and Veronica Della Dora.

Table 6.1 Map conceptions according to historical periods and authors. (After Azócar 2012)

Period	Author	Map conception
Modern Cartography	Robinson (1955)	Maps as *objective*, *scientific* representations Maps as *truths* Maps are *transparent* and ideologically *neutral*
Post-Modern Cartography	Harley (1989)	Maps as *ideologically* laden representations Maps as cultural *texts*
	Crampton (2003)	Maps as *historical products* operating within 'a certain horizon of *possibilities*'
	Casti (2005)	Maps as locus of *semiosis*; *self-referential* through iconisation
	Wood and J. Fels (2008)	Maps as *constructions* that produce the world Maps as *propositions*
	Latour (1987, 1999)	Maps as *immutable* mobiles Maps as *actants*
	della Dora (2009)	Maps as *fluid* objects, always in the making Maps as *mnemonics*
Post-Representational Cartography [a]	Pickles (2004)	Maps as *inscriptions* Maps as unstable and complex *texts*
	Kitchin and M. Dodge (2007)	Maps as *practices* (spatial practices that do work in the world) Maps as suites of cultural practices involving *actions* and *affects* Maps as *mutable* mobiles

[a] Post-representational cartography may actually be considered to be part of postmodern cartography. It is important to emphasise that Gyula Pápay also considers 'map as *images*' in the context of sciences (Pápay 2005a, b, c, Pápay 2006), but we need a further treatment and analysis for this author in another publication.

The new conceptions of maps belonging to the post-modern cartography period will be treated on the following pages. As Harley's position has already been analysed, this part will begin with Jeremy Crampton.

6.3.1 Maps as Historical Products: Horizon of Possibilities

Along this line, Jeremy Crampton draws from Heidegger's ideas when he proposes a shift from understanding cartography as a set of ontic knowledge to examining its ontological terms[1] (Crampton 2002). This means questioning the 'project' of cartography itself. Crampton also considers Edney's approach (Edney 1993) when he argues for the development of a non-progressive history of cartography: a historical ontology that, rather than being teleological, is contingent and relational.

[1] J. Crampton refers to Heidegger's work to analyse two kinds of knowledge: ontic knowledge, which concerns the knowledge of things as such, and ontological knowledge, which concerns the conditions of possibility for ontic knowledge. For more details see Crampton (2002).

For Crampton 'a historical ontology … suggests that the way things are, their being, is in fact a historical product operating within a certain *horizon of possibilities*' (original emphasis by Crampton 2002: 6). For this reason, maps are also historical products operating within a certain horizon of possibilities (Crampton 2003). This implies 'the possibility of an unfolding of the being of maps and mapping' (Crampton 2002: 13). In other words, how maps are conceptually framed in order to make sense in the world.

Therefore, maps are products of the here-and-now, no better than maps of previous generations, simply different to them. This conception differs considerably from the progressive approach of cartography, which states that maps have a constant improvement or at least development across history.

6.3.2 Maps as Locus of Semiosis: Self-Reference of the Map

Emmanuela Casti (2005) uses a semiotic approach, namely a theory of cartographic semiosis.[2] She mentions that

> A semiotic approach shifts the emphasis from maps intended as *a mediation* to maps taken as *agents*, whereby actions to be carried out on territory are determined (original emphasis by Casti 2005: 1).

Casti elaborates her ideas from the hermeneutical approach, or more specifically, from cartographic hermeneutics, which conceive the map as a tool of intervention between society and territory. In this sense, maps play an important role as agents capable of deploying self-referential information to effectively mould human intervention on territory (Casti 2005). She also deals with the concepts of *self-reference* and *iconisation*. Self-reference is by Casti defined as 'the ability of a map to be accepted as such and, at the same time, to play a role in communication that is independent of the intentions of the cartographer who produced it' (Casti 2005: 10). This statement implies that the hermeneutic approach goes beyond the deconstructive one. According to Harley, the author of a map (a cartographer) has the intention of communicating something. For Casti, however, maps with their own sets of rules produce additional meanings that affect the perception that observers have of certain places.

On the other hand, Casti defines iconisation as 'the communicative issue whereby the self-referential mechanisms of the map are used to convey conjectures as truth' (Casti 2005: 11). For her, the map itself is also an icon, namely, 'an instrument by means of which one carries out a metamorphosis of the world'. In these terms, a map is conceived as a *locus of semiosis*, self-referential through iconisation. Thus

---

[2] "Semiosis is the process whereby information is produced and transmitted" (Casti 2005: 5).

> Cartography semiotic has shown not only that maps can convey complex information but also that this information is always the product of iconization, and that it is connected with reality but cannot simply be superimposed upon it (Casti 2005: 12).

To join these two concepts, self-reference and iconisation, Casti concludes that the map as a model *replaces* rather than *represents* territory. In summary, this posture again questions the metaphor of the map as the mirror of reality: the map is not the reflection of the territory.

At this point, it should be mentioned that with her programmatic statement, Emmanuela Casti is similar to Alfred Korzybsi, who explicitly stated that 'The map is not the territory' (cited after Berque 2013: 241). It is interesting that Augustin Berque, an orientalist and philosopher (who however, also studied geography), deals with the question of whether Casti's or Korzybski's view is true in his paper "Die Transgression der Karten" ("The transgression of maps")[3] Referring to two of his earlier books (Berque 2008, 2010), he concludes that the *trajection of reality* has always and permanently been moving from the territory to the map.

6.3.3 Maps as Propositions: Perimap and Epimap

John Krygier and Denis Wood analysed maps and the discourse about maps, and pointed out that "maps are propositions". In some of their articles, they posed this conception by means of graphical schemes to facilitate its comprehension (Krygier and Wood 2009) (Fig. 6.1).

Along the lines of Kitchin (2008) and Kitchin et al. (2009), they analyse the work of Denis Wood and John Fels (2008a, b), and argue that maps produce the world by making propositions which are placed on it. Maps achieve their work by exclaiming such propositions in the form of postings of information on the maps.

An important idea of Wood and Fels is the *paramap*. All maps have an inherent authority which is conveyed by the map through the so-called paramap. This paramap 'can be broken down into a perimap element and epimap elements' (Wood and Fels 2008b: 192). The former involves the production surrounding a map, and the latter is the discourse about a map to shape its reception (for details see Wood and Fels 2008b).

For Tom Koch 'the paramap consists of the *peri*map (elements of which include ancillary maps, legends, scales, and so on) and a broadly conceived *epi*map including the article within which a map may be embedded' (Koch 2008). Koch and Denis Wood and John Fels all argue that the map image itself cannot be understood except as embedded in a paramap "that surrounds and extends a map in order to present it" (Koch 2008: 49).

---

[3] Initially written in French in November 2010 and subsequently translated into German by Marion Picker (cf. Picker et al. 2013; Picker 2013).

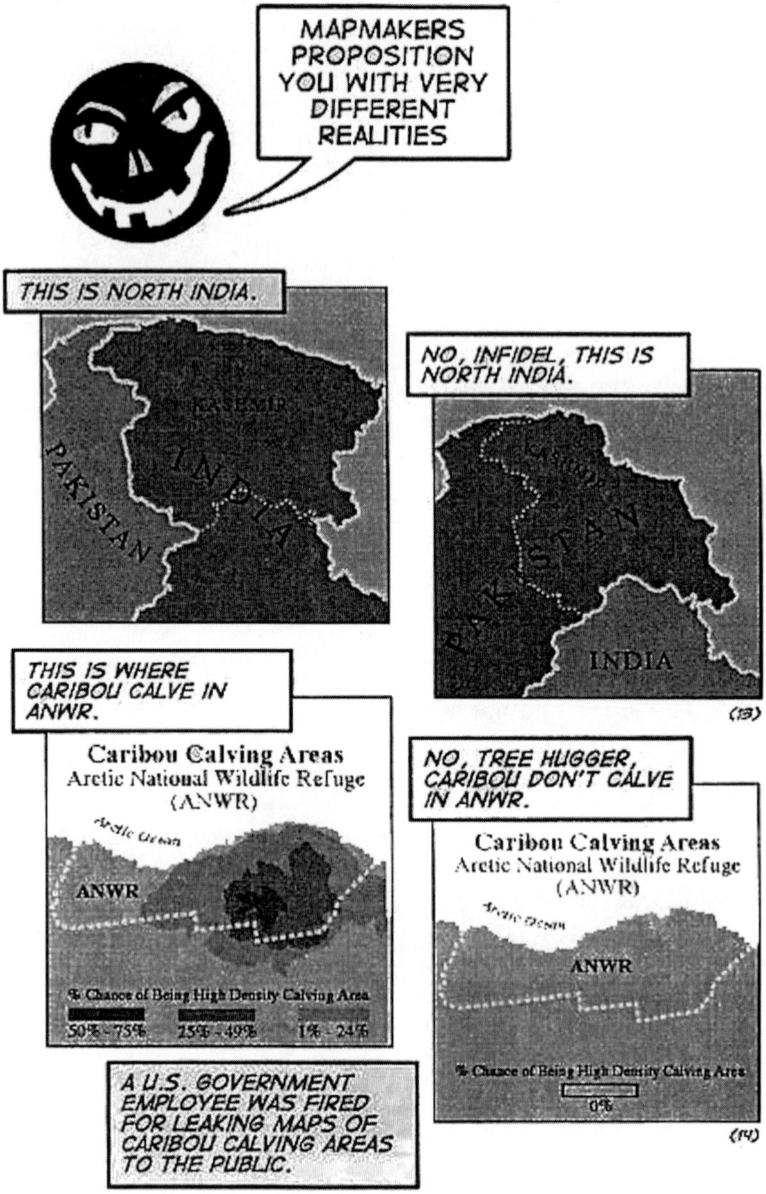

Fig. 6.1 *Ce n'est pas le monde* (After Krygier and Wood (2009: 204). In (13) each map proposes a different "Kashmir". In (14) each map proposes a different region of caribou calving in the Artic National Wildlife Refuge. (cf. Krygier and Wood 2009: 204/214).

For Wood and Fels, the authority of maps, albeit apparently descriptive, is inherently prescriptive. On the other hand, for them 'the map is nothing more than vehicle for the creation and conveying of authority about, and ultimately over, territory' (Wood and Fels 2008b: 190).

From this it can be concluded that maps are a prescriptive and not a descriptive system of propositions.[4] This proposal leads to the following:

> Given that we *see maps* as systems of propositions (as *arguments*), nothing could be further from what we have in mind. This question is *not* for us how things are arranged for the eye, but how the design promotes and constrains, how it directs, the construction of meaning. It is not about the "presentation of information." It is about the construction of meaning as a basis for action. It is for us a question of cognition (original emphasis by Wood and Fels 2008b: 194).

In this manner, these authors propose that *cognitive linguistics*[5] is an appropriate model for thinking about cartography. It means that map design should be rethought as a form of *cognitive cartographics.* In short, for Kitchin et al. (2009) this implies that employing the cognitive cartographic approach will create a non-representational approach to map design on the construction of meaning rather than graphic design and the nature of signs.

Finally, the perimap and epimap elements are related to Harley's posture about the internal and the external power of cartography respectively. According to the rules of cartography pointed out by Harley, the technical production of maps corresponds to the perimap, and the cultural production of maps is associated with the epimap elements.

6.3.4 *Maps as Immutable Mobiles: Stable Form of Knowledge*

Drawing from Latour's ideas, Kitchin et al. (2009) called maps *immutable mobiles.* Bruno Latour (1999) uses examples of cartography to show the extent to which the production of Western scientific knowledge grew in importance through time. In other words, he shows how from a historical viewpoint the cartographic practices and the mapping technologies helped Western hegemony. In this way, maps are considered immutable mobiles:

---

[4] A descriptive map only describes space, i.e. a territory, whereas a prescriptive map produces and reaffirms territory (Wood and Fels 2008a).

[5] Tom Koch mentions that Wood and Fels use cognitive linguistic as an interpretative tool. 'They propose a *cognitive cartographics* in which *mental maps* are replaced by cognitive, mental spaces as a flexible frame within which meaning is constructed. That construction is played out in the layout of the map itself' (Koch 2008: 49).

> Stable instruments for the transfer of knowledge from one place and time to another, or containers of information gathered at a specific geographic location, returned to a "centre of calculation[6]", and then plunged once more into circulation (cited by Della Dora 2009: 252).

For Latour, the scientific basis of map-making and map-use became conventionalised. Thus, maps increasingly took on the status of immutable mobiles: 'Map became a stable, combinable and transferable form of knowledge that is portable across space and time' (Kitchin et al. 2009: 18).

On the other hand, the adjective 'immutable' means, in part, that cartography theory and practices *disciplines* its practitioners and also it *silences* other local mapping knowledge, i.e. indigenous cartographic practices are considered of minor status compared to technical/professional cartography. For Latour

> The immutability, combinability and mobility of maps allowed exploration, trade and ultimately colonialism to develop by allowing control to be exerted from afar and knowledge about new territories to be effectively transported globally (Kitchin et al. 2009: 19).

This approach is consistent with Casti's claims (2005), because maps do not simply represent space at a particular time, but rather maps *produce new space-times*. In a way similar to Crampton's statements (2003), maps open up new possibilities creating new geographies and histories.

6.3.5 Maps as Actants: Social Context of the Map

Another perspective of maps that emerge in the postmodern period was that, according to Bruno Latour (cf. Kitchin et al. 2009), 'to understand maps .... it is necessary to unpick the cultures, technologies and mechanics of how a particular form of mapping came to gain immutability and mobility to reveal its contingencies and relationalities' (Kitchin et al. 2009: 15). In this way, Latour develops the Actor Network Theory (ANT) which is 'a framework for considering how maps work in concert with other actants and actors to transform the world' (Kitchin et al. 2009: 16). This network corresponds to the social context in which the map is created and developed.

Latour's contribution is that

> Maps do not have meaning or action on their own; they are part of assemblage of people, discursive processes and material things. They are deployed in an actor-network of practices rather than existing as de-corporalized, a priori, non-ideological knowledge objects (Kitchin et al. 2009: 16).

The above statement implies a philosophical and epistemological change. The emphasis is not on *what* the map represents, but on how it is produced and how it

---

[6] Key sites of cartographic practices which came to dominate the world during the Renaissance are considered "centres of calculation".

works in the world. In this way the map became an *actant* and not a *static* object. This vision about the map differs with the scientific approach in which the map is considered an isolated object inside a laboratory or workshop.

6.3.6 Mapping as Image Rhetorics

Franco Farinelli, a geographer from Bologna, Italy, postulates a 'critique of cartographic reason' (Farinelli 1996, 2009). In this context, the philologist Florent Gebaude and the geographer Veronique Maleval note that common reason believes in the *isomorphy* of the depiction process and the resulting presentation. 'A map, however, only becomes the territory by producing both a real and an imaginary space: maps are images that create worlds.' (Gebaude and Maleval 2013: 135/136). As these two authors well point out, the fact is that the map bears the territory not as a specific feature of the postmodern society, as postulated by Jean Baudrillard, but as something constitutive for the geographic science in general (Baudrillard 1981: 10).

Gebaude and Maleval write that the opinion that a map is 'no mirror image of the territory, no objective depiction of the surveyed environment, but a construct which works with conventions and symbols' goes back to *pragmatic semiotics* and *iconology* (Gebaude and Maleval 2013: 136). Here, we find links to the works of Charles Sanders Pierce and Ernst H. Gombrich (Pierce 1978; Gombrich 1984; Pape 1989). Yet, Ernst Bloch opposes to the 'reactionary' epistemology which believes in 'the mapping of an allegedly fixed entity'[7] (Bloch 1973: 875; cf. Gebaude and Maleval 2013).

Another quite interesting aspect raised by Gebaude and Maleval (2013) in stating that the postmodern offshoot of digital three-dimensional cartography using satellite-based software like Google Street, integrated navigation systems in cars, or mobile phones with navigation function are currently (2013), in a partly regressive process—due to ever more precise measuring methods, leaving the "classical" nadir perspective in favour of frontal or lateral views for the users. Thus, to some degree a return form modern road or hiking maps to medieval travel-distance maps (*Wegstreckenkarten*) can be noticed.

6.4 Foundations for a Post-representational Cartography

As summarised in Table 6.1, the two main theoretical directions of post-representational cartography evolved in the following way: maps as *inscriptions* and maps as *practices*. These development trends in cartography have to be attributed to John Pickles (2004) and Kitchin et al. (2009).

7 Bloch writes about '… die Kartenaufnahme eines angeblich fix Seienden'.

6.4.1 Maps as Inscriptions: Social Life Affected by Maps

Several other theorists in cartography have been following similar lines of enquiry to those followed by Crampton, seeking to transfer map theory from ontic knowledge to ontology (Kitchin et al. 2009). John Pickles (2004) attempted to extend Harley's observations beyond the ontic status, calling for a post-representational cartography. This perspective understands maps not as mirrors of nature, but as producers of nature. In his sense, 'A map is not a representation of the world but an inscription that does (or sometimes does not do) work in the world' (Pickles 2004: 67).[8]

Pickles proposed a shift based on the discourse of cartography up to about the 2000s. This debate had previously been focused on how the historical transformations in social life have influenced mapping techniques and map use. Instead, he now proposes to analyse the ways in which the form of cartographic reasoning affected social life.

In this way, Pickles draws attention to the necessity of a post-representational cartography, a de-ontologised cartography and denaturalised histories of cartography. He postulates a recognition that the alternative cartographic practices have an ontological status just like those developed in scientific fields.

According to Kitchin and Dodge (2007) and Kitchin et al. (2009), denaturalised histories of cartography consist of genealogies of how cartography has been naturalised and institutionalised across space and time, as a particular form of scientific practice and knowledge. Moreover, a de-ontologised cartography accepts counter mappings as having the same ontological status as scientific cartography.

6.4.2 Maps as Practices: a State of Becoming

Kitchin et al. (2009) call the period of cartography that began after the 2000s from ontology to ontogenesis. They also declare their worry about the *ontological security* of the map,[9] a preoccupation that is implicit to other authors such as Crampton, Pickles, and Latour. In other words, 'The map might be seen as diverse, rhetorical, relational, multivocal, and having effects in the world, but is nonetheless a coherent, stable product—a map' (Kitchin and Dodge 2007: 334).

---

[8] Kitchin and Dodge (2007) also draw on the work of Pickles (2004).

[9] In general terms, ontological security involves having a positive view of oneself, the world, and the future. This term moved to the field of cartography, its foundational ontology is that the world can be objectively and truthfully mapped using scientific techniques that capture and display spatial information. Therefore, the ontological security of a map is referred to the knowledge underpinning cartography and map use as learned and constantly reaffirmed (Kitchin and Dodge 2007).

Table 6.2 Map conceptions supporting post-representational cartography. Adapted from Kitchin et al. (2009). (After Azócar 2012)

Author	Map conception
Corner (1999)	Maps as *re-creations* of territory Maps and territories are *co-constructed*
Ingold (2000)	Maps as views from somewhere bound within the *practices* and *knowledge* of their makers Maps as *histories* in movements
del Casino Jr. and S. Hanna (2006)	Maps as *mobile subjects* whose meaning *emerges* through socio-spatial practices Maps as both representations and practices *simultaneously*

Nowadays this position regarding map security has been rejected by those that have adopted the performative and ontogenetic understandings of mapping. In this way:

> Maps rather are understood as always in a state of becoming; as always mapping; as simultaneously being produced *and* consumed, authored *and* read, designed *and* used, serving as a representation *and* practice; as mutually constituting map/space in a dyadic relationship (original emphases by Kitchin et al. 2009: 22).

These authors put out maps as practices and they base their analysis on four main authors: James Corner, Tim Imgold, Vincent del Casino and Stephen Hanna (see Table 6.2). In this period, cartography is considered a *relational perspective* with a constellation of ongoing processes, rather than a unified representation of reality (Kitchin et al. 2009).

James Corner (1999) argues that a territory does not precede a map, but that space becomes territory through *bounding practices* that include mapping. In this way, 'maps and territories are co-constructed' (Kitchin et al. 2009: 18). This argument is similar to that of Casti about the mechanism of territorialisation (cf. Casti 2005). Corner also suggests that cartographic research and practice needs to focus on mapping actions and mapping effects and not solely on the construction of maps *per se*.

Tim Ingold (2000) proposes an approach to mapping grounded in cultural practice (indigenous maps). He defines mapping as way-finding practices which involves the movement of people among several places inside a region. These movements are documented mobility and counter location in the Cartesian sense, i.e. these mapping practices 'do not detail location in space but histories in movement that constitute place.'(Kitchin et al. 2009: 18).

According to Vincent del Casino Jr. and Stephen Hanna, several works in postmodern cartography reproduce a series of binaries that separate maps as representations of space from spatial practices. Examples of these binaries are: representation/practice, production/consumption, and map/space. The authors suggest a way to interrogate these binaries, which are implicit in traditional and postmodern cartography. They state that maps are not only representations of a particular context, space, and time, but that maps are 'mobile subjects, infused with meaning through contested, complex, intertextual, and interrelated sets of socio-spatial practices' (del Casino Jr. and Hanna 2006: 36). They are interested in applying the methodological assumptions and processes towards thinking of a map as a space. Thus

> As such, map and mapping are both representation and practice (read: performance) simultaneously. Neither is fully inscribed with meaning as representations nor fully acted out as practices (del Casino Jr. and Hanna 2006: 36).

After examining the ontological status of maps and drawing on the aforementioned authors, Kitchin and Dodge (2007) point out that a conceptual turn in how to think about maps and cartography is significant: 'That is a shift from ontology (how things are) to ontogenesis (how things became)—from (secure) representation to (unfolding) practice' (Kitchin and Dodge 2007: 335).

Therefore, according to these authors, a map is always a result of the moment, of the context. A map only exists in *practice*. Therefore, a map is not a *product* but it always is *process*. The map *happens* or *occurs* only when someone interprets a given visual form, so it is always *practical*.

For Rob Kitchin and Martin Dodge, a spatial representation[10]

> …is rather a set of points, lines and colours that takes form as, and is understood as, a map through mapping practices (an inscription in a constant state of re-inscription). Without these practices a spatial representation is simply coloured ink on a page. … Practices based on learned knowledge and skills (re)make the ink into a map and this occurs *every time* they are engaged with – the set of points, lines and areas is recognized as a map; it is interpreted, translated and made to do work in the world. As such, maps are constantly in a state of becoming; constantly being remade (original emphasis by Kitchin and Dodge 2007: 335).

These ideas are also shared by Veronica Della Dora. Her approach is focused in the process of encountering and performing maps, more than maps as finite representations. She calls for "a re-conceptualization of maps as fluid objects that are always in the making" (Della Dora 2009: 240). For her the cartographic representations are 'mutable mobiles', i.e. maps and atlases. They are transitory and fleeting products of specific physical encounters in space and in time.

On the topic of the ontogenetic understanding of maps, and according to this new perspective, Kitchin and Dodge (2007) state that

> Maps *emerge in process* through a diverse set of practices. Given that practices are an ongoing series of events, it follows that maps are constantly in a state of becoming; they are ontogenetic (emergent) in nature. Maps have no ontological security, they are of-the-moment; transitory, fleeting, contingent, relational and context-dependent. They are never fully formed and their work is never complete. Maps are profitably theorized, not as mirrors of nature (as objective and essential truths) or as socially constructed representations, but as emergent (original emphasis by Kitchin and Dodge 2007: 340).

---

[10] Kitchin and Dodge present a thematic map concerning the "Population change in Ireland, 1996–2002" and they ask: 'Is the image a map'? (cf. Kitchin and Dodge 2007).

The statement is very important for cartography because considered as a profession, it is repositioned as a *processual* science, as opposed to a *representational* science. This means that cartographic research becomes refocused as a *science of practices, not representations*. Consequently a post-representational view of cartography is since around 2010 still emerging.

For Kitchin and Dodge

> The important question is not *what a map is* (a spatial representation or performance), nor *what a map does* (communicates spatial information), but *how the map emerges* through contingent, relational, context-embedded practices to solve relation problems (their ability to make a difference to the world); to move from *essentialist* and *constructivist cartography* to what we term *emergent cartography* (emphases added, Kitchin and Dodge 2007: 342).

The above statement means that post-representational cartography has a different perspective from that of traditional cartography and, even from historical critical cartography. In fact, from an epistemological perspective for Kitchin and Dodge,

> What this means is that science of cartography (how maps are produced) and critical analysis of cartography (the history and politics of cartography) are both positioned as processual in nature. Rather than one asking technical questions and the other ideological, both come to focus on how maps *emerge* through practices; how they come to be in the world (original emphasis by Kitchin and Dodge 2007: 342).

The authors add:

> Cartography shifts from being ontical in status, wherein the ontological assumptions about how the world can be known and measured are implicitly secure, to an ontological *project that questions* more fully the work maps do in the world (emphasis added, Kitchin and Dodge 2007: 343).

This statement, again, reflects the scope of current cartography (2013) and the changes in perspective that it contains. These new conceptions evidence a perspective shift in maps and mapping, which represents the basics for post-representational cartography, a new way of thinking in cartography.

Additionally, this perspective has effects on the philosophical and epistemological foundations of cartography. In this sense it is possible to visualise a *paradigmatic shift* in gestation that goes beyond that of an *ontical* and *ontological* approach of the discipline.

Chapter 7
Possibilities of Paradigms in Cartography

7.1 Introduction

Basically, there are two ways of analysing whether Kuhnian-type paradigms exist in cartography. First, we apply the *criteria of contrast* used in the distinction of science types. These criteria establish the differences between formal sciences and factual sciences and also distinguish between regional and quantitative geography. Second, an analysis based upon the 'tendency distribution in the epistemological-space' (Azócar 2012) is applied. The distribution or 'location' of tendencies permits the identification of paradigmatic-shifts according to epistemological and philosophical 'coordinates'. The term coordinates corresponds to the three bases of modern thought: positivism-empiricism, realism-structuralism, and idealism-hermeneutics. This approach is applied to modern geographic thought and then to the cartography of the modern and post-modern period. The aim is to locate cartographic tendencies according to the epistemological coordinates rather than to describe the technological changes that occurred during the development of the discipline.

7.2 Criteria of Contrast

In his comparison between factual and formal sciences, Mario Bunge (1998) identified criteria of contrast[1] which are important to distinguish between these two types of critical sciences. They are, among others: study object, research aims, methods and techniques, results in research or/and practice, purpose or finality. These criteria can be adapted according to different sciences and disciplines. Thus,

[1] These criteria of contrast are also called 'contrasting parameters' (reference). In this book, however, they are named criteria of contrast due to their epistemological and theoretical nature. A criterion of contrast is an indicator that permits the description of distinctions between types of sciences, e.g. formal/factual sciences or physical/social sciences. Some criteria of distinction are the following:

P. I. Azócar Fernández and M. F. Buchroithner, *Paradigms in Cartography*,
DOI: 10.1007/978-3-642-38893-4_7,

criteria of contrast are used to reach the proposed objectives to verify whether the development tendencies in cartography correspond to the paradigmatic trends experienced in the development of sciences in general.

Some criteria of contrast are the following:

- Study object: main subject (topic) analysed in each discipline or science.
- Research aims: what led to an explanation and prediction of reality through laws and generalisations or only a description of reality through unique and particular cases.
- Research method: covers the general methodology used: hypothetical-deductive or empirical-inductive or both.
- Statement type: refers to the establishment of synthetic propositions, e.g. factual sciences, or analytical ones, e.g. formal sciences.
- Purpose or finality: distinguishes between explanation and prediction of reality, i.e. natural/physical sciences, or only its description, i.e. human/social sciences.

Table 7.1 shows the comparison between factual sciences and formal sciences. A first criterion is the study object. In this case, facts and phenomena of our experience are studied by factual sciences, i.e. physics, chemistry, biology, and entities of ideal character—abstractions—are treated by the formal sciences, i.e. mathematics, logic, geometry. Certain research methods are used by the respective science type and associated with it. Then different statement types are established: On the one hand there is the empirical contrast with synthetic propositions in the factual sciences, on the other hand the logic-deductive demonstrations with analytical propositions or statements in the formal sciences (Bunge 1998).

According to the above methods, the purpose or finality is to reach the desirable knowledge. In this case, the purpose is descriptive, explicative, and predictive for the phenomena belonging to the factual sciences in comparison to purposes about the construction of the abstract thought system of the formal sciences.

Among others, the epistemology of science aims to take into account criteria for the distinction between the types of sciences or disciplines that help to increase our particular knowledge of the physical and abstract world.

Table 7.1 Criteria of contrast for factual and formal sciences based on the conception of Mario Bunge (adapted from Bunge 1998)

Factual sciences	Criteria of contrast	Formal sciences
Facts and phenomena of the experience	*Study object*	Entities of ideal character
Empirical contrasting	*Research method*	Logic-deductive demonstration
Synthetic propositions	*Statement type*	Analytical propositions
Description, explanation, and prediction of phenomena of the universe	*Purpose (finality)*	Construction of abstract systems of thought

7.3 Contrasting Paradigms in Geography: An Example

Apart from those mentioned above there are other criteria of contrast, according to the peculiarities of each discipline. In geography for instance, there exist: paradigm name, tendency or school of thought, general methods, and cartographic product (as a practical result). These criteria are useful for differentiating internal tendencies in a discipline.

Some criteria of contrast which establish the differences between two traditional tendencies in the field of geography are presented in Table 7.2. As shown in the previous sections, there was a distinct difference in the geographic thought between traditional regional geography and a new trend called quantitative geography during the first half of the twentieth century. The former is underpinned by historicism and the latter by neo-positivism. In Table 7.2 these differences are presented according to the established criteria of contrast. For traditional geography, the study object is the region (or a specific place) and for quantitative geography, the spatial relationship. For more details regarding these tendencies see Harvey (1969), Capel (1983, 1998), Gomez Mendoza et al. (1988), Ortega Valcárcel (2000).

Table 7.2 shows the general method and the more specific methods and techniques, i.e. approaches, used in geographic tendencies. Regional geography commonly applied the empirical-inductive method, and its approaches were classificatory, comparative, and historical. Quantitative geography, however, has been using the hypothetical-deductive method, and its technical approaches were mainly statistics, modelling, and data correlation. These techniques permitted the verification of previously established hypothetical statements.

The research results of regional geography led to typologies and the classification of places and regions in the world. These typologies are considered to be particular

Table 7.2 Criteria of contrast for two paradigmatic tendencies within the field of geography (after Azócar 2012)

Classical geography	Criteria of contrast	Modern geography
Historicism	*Paradigm's name*	Neo-positivism
Regional geography	*Tendency /school*	Quantitative geography
Regions, places	*Study object*	Spatial relations
Description of the directly observed physical world (region)	*Research aims*	Explanation and prediction of spatial relations
Empirical-inductive	*General method*	Hypothetical-deductive
Classificatory; comparative; historical	*Methods and techniques (approach)*	Statistics; modelling; correlations
Typologies; particular cases; unique cases	*Research results*	Generalisations, laws and theories
Monographic maps	*Cartographic products*	Statistical and correlation maps

or unique cases, meaning that the typologies are only valid for specific spatial units and not for others. On the other hand, the research results of quantitative geography led to generalisations, laws, and theories regarding the spatial relationships or spatial distribution of the phenomena under study. The idea was to apply laws and generalisations that would be valid for all spaces and places. The cartographic products used in both tendencies have also been considered as additional criterion of contrast. As a derivation of the research results, monographic maps have been produced in regional geography, whereas statistical and correlation maps have been made in quantitative geography. The latter products, especially the correlation maps, have been supported by computational and programming techniques.

To summarise: Whereas Table 7.1 establishes the differences between types of sciences, Table 7.2 indicates some criteria for the establishment of trends within a particular science or discipline. These tendencies or schools of thought can be regarded as paradigmatic shifts owing to their contrastive nature in Kuhnian terms (cf. Chap. 3). From an epistemological viewpoint, these trends also support our knowledge of a specific part of reality.

7.4 Comparing Tendencies in Cartography within the Epistemological Space

A second methodological criterion applied in our investigations is the triangular model derived from Eric Sheppard's discussion about representing critical geography and geographic information systems (GIS). These are related with the three complementary entities empiricism, realism, and idealism (cf. Sheppard 2005). In the field of cartography, Menno-Jan Kraak and Ferjan Ormeling used a triangular figure to characterise geospatial data in their three components: location, attribute, and time (see Kraak and Ormeling 2010).

This triangular model will, for our purposes, be (re)named Tendency distribution in epistemological space. Tendencies mean thoughts, trends, perspectives, and approaches which have been developed within a science or discipline, i.e. geography and cartography. Epistemological space refers to the philosophical and epistemological context that is analysed. In this way, some of the following figures depict the three main underpinnings of the modern period: positivism-empiricism, realism-structuralism, and idealism-hermeneutics. Thus, the triangular model shows how tendencies are distributed within the epistemological coordinates of modernity that frame sciences and disciplines (cf. Azócar 2012).

The paradigmatic tendencies in geography listed in Table 3.2, but this time assigned according to their epistemological space, are shown in Fig. 7.1. The positivism/empiricism coordinate depicted there comprehends three tendencies with a regular distribution: determinist geography, quantitative geography and theoretical geography (left apex). Radical geography is the only tendency along the realism/structuralism coordinates (right apex). A group of six tendencies, however, is distributed along the idealism/hermeneutics coordinates: regional

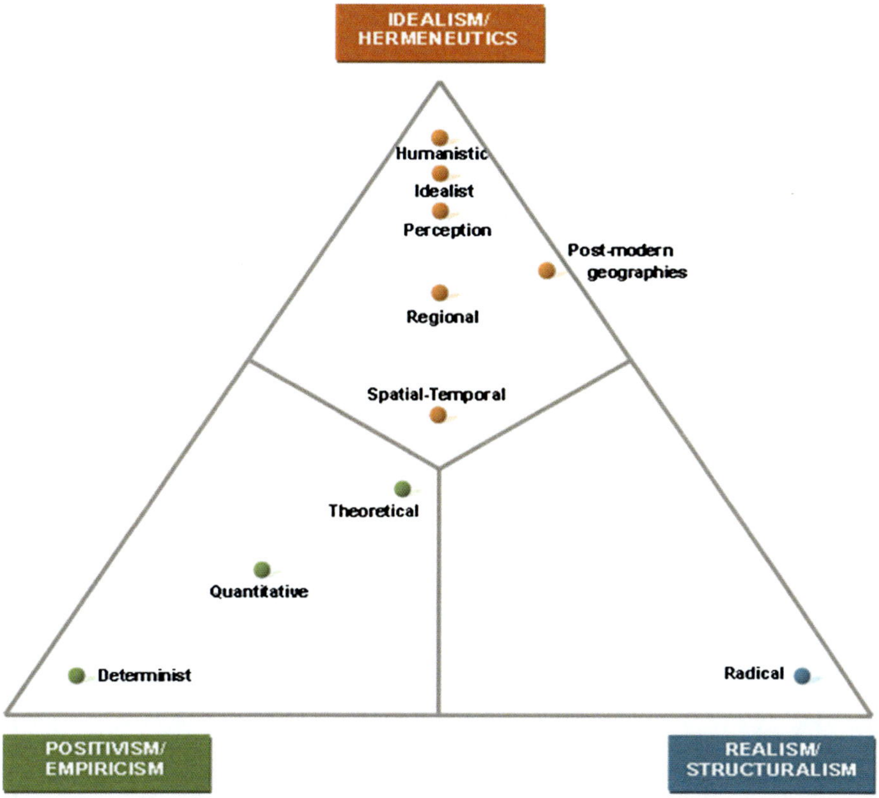

Fig. 7.1 Triangular graph of geographic paradigms (or paradigmatic tendencies) according to the epistemological bases of Modernity (after Azócar 2012)

geography, spatial–temporal geography, and the so-called post-modern geographies. Finally, in the top apex humanistic geography, idealist geography and perception geography are located in a rather concentrated pattern.

In the following, the contemporary tendencies in cartography described in the sections "Tendencies in Contemporary Cartography", "Critical Cartography in the Context of Post-Modernism", and "Post-Representational Cartography" are analysed according to the above introduced triangular model with the epistemological coordinates of the modern and post-modern period.

Figure 7.2 is a derivative of Fig. 7.1. It depicts the tendencies that were developed during the second half of the twentieth century, contemporary cartography, according to the three approaches belonging to the scientific-empirical perspective. These are the mathematical, the cognitive and the semiotic approaches. Four tendencies are, with a regular distribution, located inside this epistemological space: Analytical Cartography at the mathematical coordinate (left apex), Cartosemiotics or Semiotic Cartography at the semiotic coordinate (right apex) and Cartographic Communication at the cognitive coordinate (top apex) can be located close to the tips of the

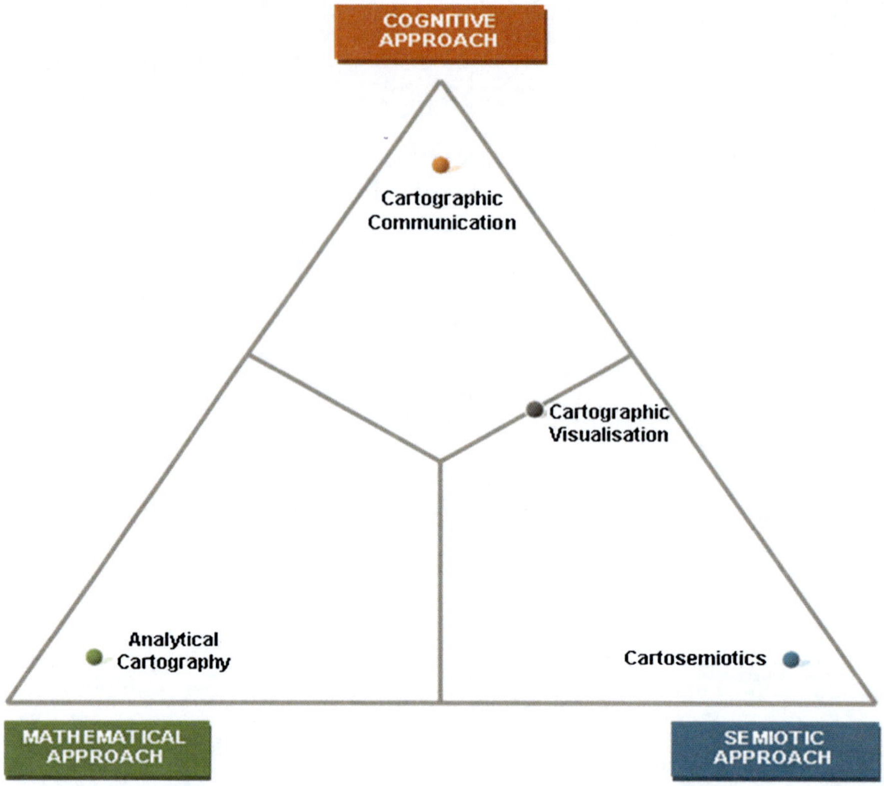

Fig. 7.2 Triangular graph showing the paradigmatic tendencies in the contemporary cartography (after Azócar 2012)

triangle. The location of Cartographic Visualisation corresponds to the cognitive-semiotic approach, according to MacEachren (1995). In this way, a clear separation of the approaches among contemporary tendencies in cartography can be depicted (Azócar 2012). This will be discussed in more detail in the following chapter.

Figure 7.3 integrates aspects of Figs. 7.1 and 7.2. It shares the epistemological space with Fig. 7.1, however, this time the structuralism axis is replaced by the post-structuralism coordinate. Critical Cartography which pertains during the post-modern period is included. The scientific-empirical perspective belonging to the positivism-empiricism coordinate of Fig. 7.2 has been considered. This perspective contains three approaches: mathematical, cognitive, and semiotic. Thus, the contemporary tendencies in cartography are situated in this part of the triangular model (left apex). On the right side of Fig. 7.3 are, in a concentrated pattern, whose tendencies located which belong to Critical Cartography.[2] Social

[2] This tendency was analysed in the section "Critical Cartography in the Context of Post-Modernism".

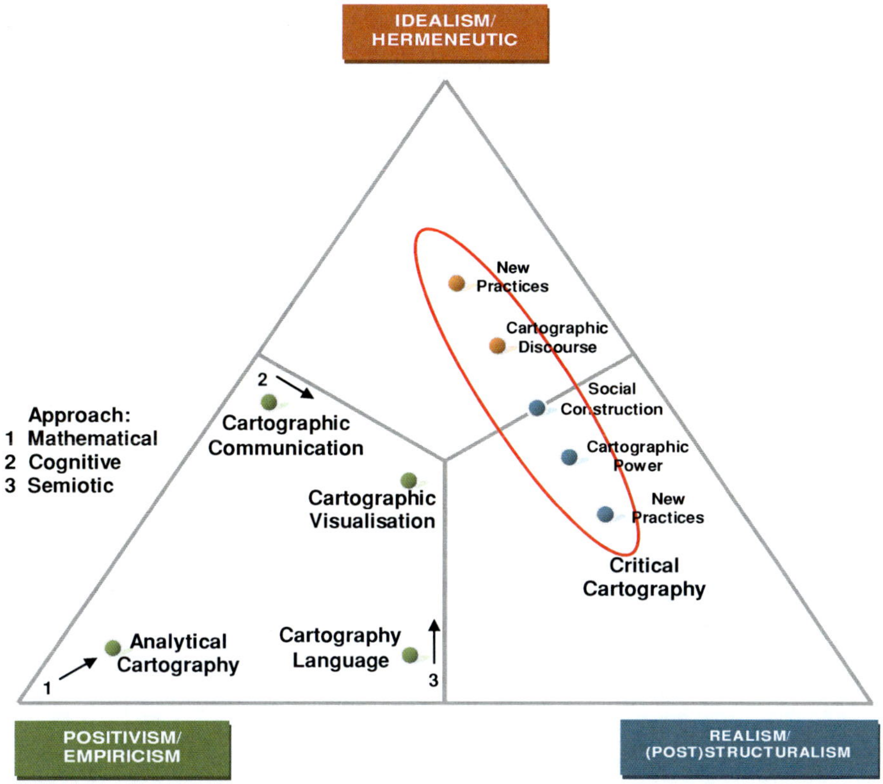

Fig. 7.3 Epistemological space model of tendencies in cartography during Modernism and Post-Modernism using the scientific-empirical and critical approaches (after Azócar 2012)

construction is shown sharing both the idealism/hermeneutic and the realism/post-structuralism spaces. Next to the social construction, cartographic discourse is located along the idealism/hermeneutics coordinate and cartographic power along the realism/post-structuralism axis. In the same way, new (cartographic) practices are considered sharing both the post-structuralism and the hermeneutics aspects.

A concept similar to that of Figs. 7.1 and 7.3 has been applied in Fig. 7.4. There, however, the three philosophical-epistemological bases of the modern period have been replaced by the ontological security[3] of maps according to Kitchin (2008). This implies a sequence of ontic-ontological-emergent levels.

[3] Ontological security is a stable mental state derived from a sense of continuity in regard to the events in one's life. Giddens (1991) refers to ontological security as a sense of order and continuity in regard to an individual's experiences. 'Confidence or trust that the natural and social worlds are as they appear to be, including the basic existential parameters of self and social identity' (Giddens 1993: 374–377). We consider ontological security as applied to cartography according to Rob Kitchin (2008).

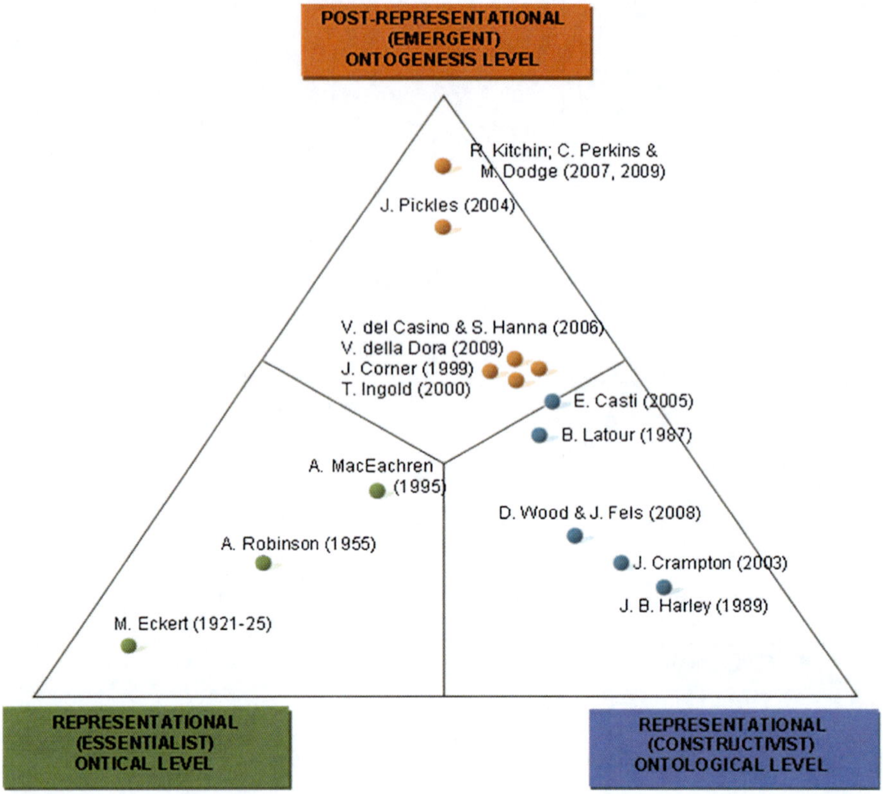

Fig. 7.4 Epistemological space model of tendencies in cartography during the Modern and Post-Modern Periods making use of representational and post-representational approaches (after Azócar 2012)

Thus, there is an ontic level which is essentialist within the positivism/empiricism coordinates (left apex of Fig. 7.4). The ontological level which is focused on a constructivist perspective is, similar to the post-structuralism coordinate, located at the right apex. The ontogenesis level which pertains to the emergent coordinate (Pickles 2004; Kitchin and Dodge 2007; Kitchin et al. 2009) is located at the top apex. According to these authors, the Representational Cartography approach would belong to both, the ontic and the ontological coordinates. The new map conceptions that arose during the Post-Representational Cartography,[4] would belong to the ontogenesis coordinates.

Figure 7.4 also shows that Max Eckert's scientific tradition, Arthur Robinson's cartographic communication and Alan MacEachren's cartographic visualisation share the ontical and essentialist coordinates (left apex). This is fostered by the

[4] This approach was analysed in Chapter "Post-Representational Cartography".

philosophical, ontological, and epistemological underpinnings of the scientific-empirical approach.

Furthermore, Fig. 7.4 summarises stances taken by authors who draw from social theory and human geography which have been applied to cartography and mapping (right apex). This is the case for John B. Harley's legacy and a group of his followers who belong to critical cartography. In this way, Harley, Jeremy Crampton, and Denis Wood and John Fels are in a concentrated pattern, located at the centre of the ontical-ontological coordinates. Similarly, there are also some authors in the critical studies context that can be classified in a transitional stage between the ontical-ontological and ontological-ontogenesis approaches (e.g. Emanuela Casti and Bruno Latour). In this transitional space, a group of thinkers including James Corner, Tim Ingold, Vincent Del Casino and Stephen Hanna, and Veronica Della Dora are also located (cf. right side of figure). They all occupy an intermediate position between representational and post-representational cartography, displayed in a concentrated pattern.

Some other authors consider cartography with a post-representational stance: especially John Pickles who is followed by Rob Kitchin, Chris Perkins, and Martin Dodge (cf. top apex of Fig. 7.4). Within the epistemological space model, they form a different grouping with respect to other stances. The proposals of these thinkers are under an *emergent knowledge* approach in the ontological-ontogenesis transition coordinates.

However, it should be mentioned explicitly that today in cartography, and in particular in geoinformatics, the term ontology is frequently applied in a different way than initially used in analytical philosophy. This is, in part, due to misinterpretations in publications of the early years of the 21st century, like Kitchin et al. (2009). According to Zsolt Török (courtesy written communication 2012) we want to emphasize that ontology must not be explained in terms of epistemology. Such a wrongly materialised interpretation leads to a problematic concept of representation in science, so to the concept of science as 'the mirror of nature'. By 1989, i.e. 12 years before Crampton (2001, 2002), Török suggested *historical social ontology* as the theoretical basis for a 'new' cartography (Török 1989). This approach is still open to the community interested in theoretical cartography research.

To conclude, using the triangular graph of Fig. 7.4 it is possible to locate the different cartographic tendencies within the epistemological space of contemporary cartography and post-modern period. The distribution of these tendencies inside the figure, i.e. their concentration-dispersion pattern, permits the identification of some paradigmatic shifts (or their absence), according to the scientific communities of the Kuhnian terminology. This aspect will be deepened in the next chapter.

7.5 New and not so New Epistemological Crises in Cartography

To mention it right away and explicitly: The following deliberations draw extensively on an article by the German scientist Marion Picker in a book co-authored by her (Picker et al. 2013) where she as a studied media scientist, anglicist, germanist *and* philosopher quite profoundly—however logically from the viewpoint of an expert in humanities—dealt with paradigmatic changes in cartography.

Almost 90 years after the seminal statements of Max Eckert 'to establish theoretical cartography as a science',[5] in his article about the future of cartography Alexander Woldtschenko casually asked the question whether cartography is *still* a science (Wolodtschenko 2009).[6] The implicit answer is, though: yes; but for the remainder several reasons are given why scientific cartography runs the risk to become irrelevant, i.e. due to its 'methodical-conceptional deficits' which have to be assigned to the predominance of applications, applicabilities and technologies over theory (Wolodtschenko 2009: 45; cf. Koch 2004: 5). Already Max Eckert admonished not to leave cartography to the 'technologists' (Eckert 1921: 2; cf. also Glasze 2009: 181, 187). Eckert's oevre per se, however, demonstrates how far future-oriented questions are just emerging from the interpenetration of technological, historical-cultural and epistemological aspects (Picker 2013: 8).

Other than Wolodtschenko (2009), Denis Wood located the "problem" of cartography at those who consider themselves as the scientific guardians of cartography. Drawing on the famous French formula[7] and a low pun by Friedrich Nietzsche, in his seminal paper "Cartography is Dead (Thank God)!" he welcomes the end of the exclusive pretensions of the academic and merely professional cartography to the relevant map knowledge. Wood claims that geoinformation systems and their successful commercialization actually lead to a redefinition of cartographic competence. Despite Wood's commitment to the "classical" principles cartographic technology, he is rather inclined to vote for a high-quality "map making" (Wood 2003: 6). Wood appreciates a simply scribbled sketch map not less than a map which satisfies all scientific requirements. Together with John Krygier he even issued a Web-based sort of tutorial or manual: http://makingmaps.net/ (cf. Krygier and Wood 2005; Picker 2013: 8).

Not only according to Picker (2013) several authors sense a certain crisis of cartography at the beginning of the 21st century (cf. amongst others Hruby and Guerrero 2008: 1, 7 and Crampton 2010: 4). *However*, there exists general consent that the "age of space" Foucault (2005: 931) has not yet begun to cease, and hence also the generation, use, thematisation, historisation and re-invention of maps. Marion Picker (2013: 9) tried to investigate into the amazing relation between the

---

[5] "... die theoretische Kartographie als Wissenschaft zu begründen ..." (Eckert 1921: III, drawing on Eckert 1907).

[6] "Ist die Kartographie noch eine Wissenschaft?" (Wolodtschenko 2009: 58).

[7] 'Le roi est mort, vive le roi!'.

still exiting popularity of maps and the asserted or supposed crisis of cartography. Stating that while during the 1960s in cartography a structuralistic, semiologic and communication-scientific change occurred (cf. Bertin 1967 and Freitag 1980, 2001), she sees at the beginning of the 21st century competitive situations coming up due to the various extra-scientific developments. The possibilities of *Web 2.0 Cartography* caused specialized knowledge that is now maintained by disciplines like geoinformatics or (geo)visual analytics, thus claiming even parts from the core of the research area of traditional cartography (cf. Hruby and Guerrero 2008: 9).

It is interesting that a new conceptualization of cartography takes its motivation exactly from what is by Picker (2013: 9) considered its crisis. It develops, like in Jeremy Crampton's book, a critique of both map and cartography, recently also including "critical GIS studies" (Crampton 2010). As also pointed out by Marion Picker, this *critical cartography* that developed in the Anglo-American world since the 1990s, in particular due to the reception of Brian Harley's publications, not only investigates the discursive conditions of map production in and on maps but also maps as models and programmes for social and political processes (cf. Picker 2013: 10). This critical attitude has by no means being a matter of academic concern only. According to Cosgrove (2005: 27) the diversity of cartography has primarily to be attributed to the *cultural turn* occurring, one way or the other, in almost all human and social sciences within the aforementioned time. In this context maps are considered as diversified practices within their historical changes (Picker 2013: 10).

As mentioned elsewhere in this book (cf. Sect. 5.5), in the first volume of the multi-volume *History of Cartography* Brian Harley and David Woodward take these developments into account by defining maps in a general, rather descriptive than prescriptive manner in the following formulation: 'Map are graphic representations that facilitate a spatial understanding of things, concepts, conditions, processes, or events in the human world'[8] (Harley and Woodward 1987: xvi). Denis Cosgrove, who has actually also been in favour of the term *mapping(s)* (Cosgrove 1999, 2005), attribute much of the "*fashionable fascination*" of maps to this epoch-making work *History of Cartography.*[9] Trying to differentiate *mappings* from *maps* (which he does not accomplish very rigorously) he turns against various fields of cognition science that, through terms like *cognitive map*, *mind map*, *concept* or *semantic map*, decisively contributed to make alternative conceptual characteristics of maps known and to circulate them (Cosgrove 1999: 3). In his 1999 publication book he points out that in the end every mapping process is necessarily also a cognitive one (Cosgrove 1999: 7) and, hence, cognitive maps are

[8] To the authors' opinion this definition certainly hast o be extended by the *natural world* as well as *other celestial bodies*.

[9] Since for Cosgrove the graphic and representative qualities are crucial criteria of a map's nature ("*Kartizität*" Picker 2013: 12), he also includes narrative, literary route descriptions but also abstract geometric compositions in his far-reaching/broad/wide/widely formulated/broadly formulated definition (Cosgrove 1999: 1, 17).

subject to determinative discursive conditions. In a 1996 paper in the map-historical Journal *Imago Mundi* Matthew Edney writes:

> One of my favourite instances of how the idea of the map is accepted automatically is provided by educational and psychological theorists in the United States. Numerous scholars have written extensively on what they refer to as 'semantic mapping' [...] Despite the evident importance of the map as a metaphor for the concept, their literature seems to contain no discussion of what 'mapping' is. The sub-discipline as a whole is underpinned by an unexamined assertion, and authors rely on the communal understanding of what maps are. (Edney 1996: 187).

If, as Picker (2013: 14) sets forth, *cartography* has traditionally been defined as the technique of map making, the knowledge about and history of maps as well as their entirety, and *maps* as cartography's object, at the turn of the 2000s to the 2010s both concepts are converging. The insecurity about the decisive criteria of the nature of a *map* is increased by the new technological possibilities for the production, storage, communicability, and visualisation of spatial data. As the statements by Harley and Woodward (1987) as well as Cosgrove (2005) showed, all parameters that traditionally played a role in map definition, like format and material, extension of depicted region, the map's "visibility", its geographic names and lettering in general, the overall design, scale (be it numerical or graphic), map frame, geometric grid etc. gained new terminological "fuzzinesses" and were called "instable" by Picker (2013: 15).

Credits belong to Marion Picker that, based on her humanities background, she managed to link the above statements with the psychoanalytical aspects which—surprise, surprise!—have some relation to maps (cf. Picker 2013: 15, 16). It is by any means remarkable that Sigmund Freud, when discussing a possible reconciliation with *culture*, the main point in his publication '*Die Zukunft einer Illusion*' ('*The Future of an Illusion*'), used a geographic-cartographic example to describe a type of knowledge resp. science that is *without illusion*, i.e. not determined by desires and fears. Freud claims that, in contrast to religious dogmas, for geographic theorems it is possible and sufficient to make a personal check. A simple journey, so Freud, suffices to verify the tenets depicted in a map. He gave the example "Constance is situated at Lake Constance",[10] a saying that is since Freud's use well-known in the German-speaking region. Using the statement "The beautiful town is located at the banks of a large body of water that all local residents call Lake Constance"[11] he discloses, however, at least indirectly, that the principle of self-verification/individual checking can in many cases be problematic, in particular in cartography. The pivotal criterion is based upon hearsay or second-hand knowledge: what seems to be natural, is merely denominative convention; a finding which may be known to every mapping cartographer who ever did her/his work in unfamiliar remote areas where no reliable maps exist (cf. Buchroithner 2011: 24, 25, 31).

[10] "Konstanz liegt am Bodensee." (Freud 1974: 159).

[11] "Die schöne Stadt liegt am Ufer eines weiten Gewässers, das alle Umwohnenden Bodensee heißen." (Freud 1974: 159).

7.6 Autostereoscopic True-3D Cartography: Another New Paradigm

Immediately before and soon after the turn of the millennium, a series of papers initiated and authored by a cartographic 3D-visualisation group at the Dresden University of Technology, Germany, indicated another paradigmatic development in cartography: the advent of the third dimension in the form of stereoscopic perception using physically flat lenticular foil maps (Buchroithner 1999; Buchroithner and Schenkel 2001; Buchroithner et al. 2000, 2004a, b, 2005a, b, c). These flat maps allow the *spontaneous* stereoscopic vision *without any viewing aids*: This is called *autostereoscopy*. Much of the knowledge gained through these developments, which were unfortunately a little bit unnoticed by the global cartographic community, has been summarised in a monograph by Thomas Gründemann (Gründemann 2004) and then, in a comprehensive form, been published in a seminal journal paper entitled "True Three-Dimensionality in Cartography: Yesterday, Today and Tomorrow" ("Echtdreidimensionalität in der Kartographie: Gestern, heute und morgen"; Buchroithner 2007).

There were, however, precedent attempts made to introduce true-3D cartography by means of autostereoscopy, in these cases by means of holography: at the 19th International Cartographic Conference in Ottawa in 1999 the first actual holographic map[12] was presented to the global cartographic community. This whitelight hologram even allowed the viewing, in a flip mode, of the unlabelled 'pure' landscape or the landscape plus the geographic name tags hovering above the terrain and thus (cf. also oral communication by Harold Moellering 1999) represents the first actual holographic map (Buchroithner and Schenkel 1999; Kirschenbauer and Buchroithner 1999; Buchroithner 2000). The extensively high production costs for both 'simple' holographic stereogrammes as well as 'rea' cartographic holograms finally abandoned further developments into this direction for a while (see next paragraph) and fostered the 'triumph' of lenticular foil technology (Buchroithner and Knust 2013a). A further enhancement of this new trend was reached by displaying three different scales in a three-in-one flip mode 'flying carpet approach' where part of the depicted landscape is represented at a large scale in smaller maps hovering above the terrain (Buchroithner et al. 2005a, Buchroithner 2008). The multitemporal domain was also added to this true-3D cartography by displaying two or even three instants of time in one lenticular foil map in a stereo-flip mode (El Nabbout 2007; Bruhm et al. 2010).

Within the first decade of the 21st century, however, holographic relief-map production got a new boost in the non-civilian domain. Apart from a few groups in

[12] By, 1985 the Canadian R. Simard from the Canada Center for Remote Sensing (CCRS), together with the MIT Professor of Media Arts and Sciences, Stephen A. Benton (1941–2003), produced an achromatic holographic stereogram out of Landsat multispectral scanner data. This, however, was only a sort of stereoscopic image without any further cartographic information like geodetic grid or name labellings.

North America, a young team in Turkey has been pushing these developments. Thus a revival of actual holographic maps may be expected in the years to come (cf. Dalkiran et al. 2009, 2012). Whether this technology may lead to another real shift in the paradigms of cartography in the long run, due to the direction-independent viewing of "virtual landscapes", remains to be seen.

Parallel to the 'hardcopy approach', digital autostereoscopic display technologies have also been used for cartographic purposes ('softcopy stereoscopy'; Buchroithner 1998; Buchroithner et al. 2000): In 2001, at the 20th International Cartographic Conference in Beijing, conception and prototypic realisation of an interactive true-3D atlas using an autostereoscopic display developed at TU Dresden, Germany, (Liehmann 2003) was presented to the global cartographic community. In the appraisal of it during the best map awards ceremony, this has been described as another quantum leap in cartography—at least regarding the representation of the third dimension.

Although the use of autostereoscopic electronic displays has mainly been demonstrated with georelief, thematic phenomena were also the subject of research and development using autostereoscopic displays after the year 2010 (Bröhmer et al. 2012). They proved to have a significant potential, in particular for educational purposes (Dickmann et al. 2012).

Modern Web technologies also allow for the long-distance real-time distribution of truly three-dimensional geodata (Buchroithner et al. 2012; Sanchez 2012). Even smartphones with touch-screen functions can, since 2010, diplay geodata in general and maps in particular in a lenticular foil-based way in an autostereoscopic way (Buchroithner 2011).

All these developments were also accompanied by profound theoretical considerations in various ways, which are spread over the aforementioned publications (primarily Buchroithner 1999; Buchroithner and Schenkel 2001; Buchroithner et al. 2004b, Weiss and Buchroithner 2012), and amongst others that also lead to Buchroithner's postulate to extend the semiological system of Jacques Bertin into the third dimension (cf. i.a. Buchroithner and Böhm 1998; Buchroithner et al. 2004b). In hindsight, it seems surprising that so far, it has hardly been stated in the disciplinary literature that the above mentioned technological developments actually also triggered another notable paradigmatic change in cartography.

The theoretical framework and basis for that in terms of physiology, psychology and technology is given by Buchroithner and Knust (2013b). In their account about recent and future developments concerning the representation of the third dimension in maps Buchroithner and Knust (2013c) credibly prove that, like in everyday life, also in cartography autostereoscopic displays will increasingly penetrate data visualisation along the z axis.

Chapter 8
Scopes of Paradigms in Cartography

8.1 Introduction

This chapter discusses the results obtained through the use of the criteria of contrast and the tendency distribution in the epistemological space, which were applied to the cartographic tendencies treated in the previous chapter "Possibilities of paradigms in cartography".

First, some paradigmatic tendencies in cartography are analysed according to the criteria of contrast. The denomination of explicit and implicit paradigms and tendencies made by various authors during contemporary cartographic development is also investigated. Second, the transition from the scientific-empirical period to critical cartography is explained through the discussion of the works of representative thinkers of modernism and post-modernism. Here, the body of knowledge in the discipline is highlighted. Subsequently, post-representational cartography is proposed as a paradigmatic trend (in ontological terms) in comparison to traditional representational cartography. Then an attempt to categorise cartographic tendencies is made, considering three epistemological levels: scientific (essentialist), sociological (constructed), and ontological (emergent).

As a result, the paradigmatic cartographic tendencies, the scientific-empirical and critical paradigms, and the representational and post-representational paradigms in cartography are proposed (cf. Azócar 2012). Finally, a return to Kuhn's thought is used to analyse the scope and limitations for cartography's own paradigms: In Kuhnian terms, cartography is a multi-disciplinary field with several internal paradigms, but its consolidation through a real revolutionary change depends on the persuasive power of the cartographic community in its social-scientific context.

P. I. Azócar Fernández and M. F. Buchroithner, *Paradigms in Cartography*,
DOI: 10.1007/978-3-642-38893-4_8, © Springer-Verlag Berlin Heidelberg 2014

8.2 Paradigmatic Tendencies as in Other Sciences (Geography)

8.2.1 Proposed Paradigm Tendencies in Cartography

Cartography is considered as a relatively new science in comparison to the physical and social or human sciences. Nevertheless, cartography has a solid theoretical framework in the context of sciences, and can also develop its own paradigm shifts. To verify whether the development of these tendencies in cartography correspond to the paradigmatic trends experienced in the development of modernism and post-modernism, the so-called criteria of contrast are used. The criteria considered in our book are—as already mentioned in Chap. 7—study object, research aims, methods and techniques (approaches), research results, and cartographic product results (cf. Table 8.1).

Derived from a literature review of contemporary cartographic development, six tendencies are proposed and then applied to the criteria of contrast: Cartographic Language (i.e. Cartosemiotics), Cartographic Communication, Analytical Cartography (i.e. Cartographic Modelling), Cartographic Visualisation, Critical Cartography and the so-called Post-Representational Cartography (cf. Azócar 2012; cf. Table 8.1).

Table 8.1 shows that different study objects are established for each tendency. For example, map language/map symbolism, map image/map design, map model, map-use, map content, and the map proper. Along this line, different research aims target each tendency. Nevertheless, there are some similarities between Cartographic Language and Cartographic Communication: in both cases the aims are about rules and generalisation. But in the former, they are used to create symbolism in map language and in the latter, map design.

Regarding methods and techniques, the majority of tendencies uses systematic research methodologies inherent to sciences and scholarly disciplines. Table 8.1 highlights their approaches and their combinations. For instance, in Cartographic Language we have the linguistic/semiotic approach, and in Cartographic Visualisation, the cognitive/semiotics one. The hermeneutic-deconstructivist and ethnographical-processual approach in Critical and Post-Representational Cartography, respectively, are also considered. According to the nature of the maps, the integration of approaches, rather than isolated ones, is common. In other words, each tendency uses different approaches according to its research aims and study objects. This means that the tendencies in cartography have different worldviews—in Kuhnian terms—in which distinct study objects lead to different methods and approaches.

Consequently, these different approaches lead to specific research results considering maps as study objects. For instance, research results can lead to 'representational and conceptual models of the real world' in the digital environment of Analytical Cartography—in contrast to the' visual thinking and visual communication mapping' of Cartographic Visualisation; or to research results with

Table 8.1 Tendencies in contemporary cartography analysed by the criteria of contrast (after Azócar 2012)

Criteria of contrast	Tendencies					
	Cartographic language	Cartographic communication	Analytical cartography	Cartographic visualisation	Critical cartography	Post-representational cartography
Study object	Map language/ symbolism	Map image/design	Map model	Map-use space	Map content	Map *per se*
Research AIMS	Rules and generalisations in map language	Rules and generalisations in cartographic communication	Analytical modelling and hypothesis testing of mapped phenomena	How map work as visualisation tools	Uncover the power-knowledge practices embedded in maps	How maps emerge through cultural, social and spatial practices
Methods and techniques (approaches)	Linguistic-semiotic	Perceptual/ cognitive (psycho-physical)	Analytical/ mathematical	Cognitive/semiotic	Hermeneutic-deconstructivist	Ethnographical—processual
Research results	Graphic mode for expression of geospatially related data	Cognitive map-design and map use (map reading)	Representational and conceptual model of the "real" world	Visual thinking and visual communication mapping	Maps as social constructions, and instruments that exert power	Maps in a state of becoming (in process) only existing in the practice
Cartographic product results	Cartographic alphabet, grammar, reading, and writing	Functional and optimal map effectiveness	Virtual map with deep and surface spatial structures and data levels	Synthesis/ presentation maps and exploration/ analysis maps	Historical devices/ artefacts, context-dependent	Emergent maps

an internal scope: 'graphic mode for expression of geospatially related data' of Cartographic Language, in contrast to the results of a more extended ambit: 'maps as social construction and power instruments' of Critical Cartography.

As shown in Table 8.1, each tendency yields specific cartographic products like 'cartographic alphabet, grammar, reading, and writing' in Cartographic Language, and 'virtual map with spatial deep and surface structural and data levels' in Analytical Cartography. There are pragmatic cartographic products such as' functional and optimal maps effectiveness' in Cartographic Communication, and 'synthesis-presentation maps and exploratory-analysis map' in Cartographic Visualisation. There exist also controversial ones like the 'historical devices/ artefacts context-dependent' in Critical Cartography and 'emergent maps' in Post-Representational Cartography.

In summary, the criteria of contrast used in the differentiation of formal and factual sciences and internal paradigm shifts in geography (cf. previous chapter) are also useful for the identification of some paradigmatic tendencies in cartography. Each tendency has particular features that distinguish it from the others. The six tendencies developed since the 1950s until the present (2013) show specific differences some of which can be considered as paradigms of their own. According to what has been discussed in this book, cartography has a solid body of knowledge, just as other scientific disciplines such as geography do. In turn, the current tendencies can be considered to be paradigmatic stances within the cartographic discipline (Azócar 2012).

Next is an analysis of the tendencies in cartography according to some of the main authors. There are authors who belong to or herald a specific trend, e.g. A. MacEachren for Cartographic Visualisation and D. Fraser Taylor for Cybercartography) and others who analysed certain cartographic developments, such as Peterson (2002), Ramirez (2004), Cauvin et al. (2010) and others. Authors of German-speaking origins are not considered in this table because they will be specifically analysed in another publication.

Most authors in cartographic literature consider Cartographic Communication to be an explicit paradigm in the discipline. They also mention A.H. Robinson's legacy as a cornerstone of the discipline. In other words, Cartographic Communication is the most important and traditional issue in the development of cartography. Cartographic Language is probably associated with the former. For this reason it is called a tendency only by Ramirez (2004) and Cauvin et al. (2010). Nevertheless, the cartosemiotic body of literature is quite considerable especially in Eastern Europe and Russia (see Schlichtmann 1999; Wolodtschenko 2011; also Commission on Theoretical Cartography, International Cartographic Association ICA). In the first decade of the 21st century Cartographic Language has been proposed as a new paradigm in the context of hermeneutics for the stylistic diversity in topographic maps by Alexander Kent and Peter Vujakovic (cf. Kent and Vujakovic 2011).

Analytical Cartography and Cartographic Visualisation are strong tendencies that have been accepted as paradigmatic shifts by Peterson (2002), Ormeling (2007), and Cauvin et al. (2010). Ramirez (2004), however, only considers them to be cartographic tendencies. The above mentioned analytical traditions are classified as paradigmatic by Sui and Holt (2008). Furthermore, MacEachren (1995) is considered as a one of the main exponents of Cartographic Visualisation in the 1990s.

In the same sense, associated with visualisation, Cybercartography is presented as a paradigm proper by Taylor (2005). Peterson (2002) also considers Cybercartography as a consolidated paradigm. Nevertheless, Hruby (2011) questions such claims arguing that Cybercartography has not the required characteristics according to Kuhn's paradigm concept. We agree with Hruby's critique in the sense that, if criteria of contrast like study object and cartographic products results are applied, this would mean that cybermaps and cybercartographic atlases would be obtained. In this case, the object of study seems to be confused with the results.

In epistemological terms, the above tendencies fall within the neo-positivist approach. The map conception conceived in Cartographic Communication, Cartographic Language, Analytical Tradition, and Cartographic Visualisation belongs to World 3 of the Popper's Three Worlds Theory. The map as a physical device pertains to World 1, but its contents and interaction with the mind of the user, i.e. map-mental image, map model, map language, are located along the line of works of art, science, and technology, and the human language realm. Since all of these are according to Popper a product of the human mind, they belong to World 3 (cf. Buchroithner 1997).

Peterson (2002) and Sui and Holt (2008) consider critical cartography as an explicit paradigm which challenges the scientific character of the discipline (as mentioned in Chap. 4). Whereas it is an important trend for Cauvin et al. (2010) and Kitchin et al. (2007), it is a starting point from which to formulate their critique of traditional cartography. Of all the authors mentioned, only Kitchin et al. (2007) consider the so-called post-representational cartography as a possible paradigmatic shift in cartography and mapping. This is an interesting intellectual but that has not been mentioned by other authors in recent papers, such as Cauvin et al. (2010) whose historical review reaches up to the first decade of the twentieth-first century.

In summary, the six cartographic tendencies are proposed as explicit and implicit paradigms, and in some cases, only as tendencies. Peterson (2002), Sui and Holt (2008), and Taylor (2005) mention an *explicit paradigm* in cartography, whereas Ormeling (2007) and Cauvin et al. (2010) point out trends that can be considered to be *implicit paradigms*. Only Ramirez (2004) sets up *explicit tendencies*. The most important aspect is that the adjective paradigmatic, given to the cartographic tendencies (Table 8.1) by some thinkers in cartography, is independent of the tendency's features. In other words—as mentioned above—each tendency has its own features that make(s) it different from the others, independent of whether it is or is not paradigmatic.

It is important to emphasise that several German-speaking authors such as Erik Arnberger; Günter Hake, Dietmar Grünreich and Liqiu Meng; and Gyula Pápay (Arnberger 1970; Hake et al. 2002; Pápay 2005a, b, c, 2009) do not explicitly mention paradigm shifts in cartography, but they belong to the empirical and scientific tradition in the discipline. Similarly Rudi Ogrissek, Wolf Koch, and Ulrich Freitag made important contributions to the theoretical field of cartography (Ogrissek 1987; Koch 1995, 2002, 2004; Freitag 1980, 1985, 1991, 1992a, b, 2001). The definition of conceptualisation in cartography and geomatics builds upon important contributions from Jürgen Bollmann and Wolf Koch (see Bollmann and Koch 2001, 2002).

Nevertheless, Max Eckert [actually Max Eckert-Greifendorff] (10. 4. 1868 – 26. 12. 1938), German cartographer and geographer, was the most important theoretician in the field of cartography. His contributions such as "Die Kartographie als Wissenschaft" (1907), "Die Kartenprojektion" (1910) and, "Die Kartenwissenschaft" (1921, 1925) permitted to configurate the discipline as a distinct science. Eckert strove as early as 1907 for the recognition of cartography as a new science with his presentation at the Nuremberg Geographers' Day "Die wissenschaftliche Kartographie im Universitätsunterricht" (Scientific Cartography in University Education) and his article "Die Kartographie als Wissenschaft" (Cartography as [a] Science) in the magazine of the Society for Geography of Berlin (cf. Koch 2010). Also, Eckert presented his thoughts on "Map Logic" (Kanakubo 1990).

According to Eckert, the main objective of cartography is the examination of the creation process of a map as well as the analysis of its essence, structure, and possibilities of use. 'He stresses that the science of a map—Kartenwissenschaft—is sister and indispensable assistance to geography and the map itself is a philosophical stone and the eyes of geography' (Ostrowski 2008: 268). Considering these aspects, we can claim that the scientific-empirical approach in the discipline was initiated through the contributions made by Eckert. From there the literature began to include the concept of science in cartography and mapping. Also, Eckert's work initiated the writing of textbooks about cartography.

8.2.2 Distinction Between Scientific-Empirical and Critical Approaches

Matthew Edney established a separation between the *empiricist paradigm* and the *critical paradigm* in the historical development of the discipline (Edney 2007). We agree with Edney's proposal and further implication that cartographic language, cartographic communication, analytical cartography, and cartographic visualisation constitute the empiricist paradigm he posed. Additionally, within this perspective (shown in Fig. 7.2 of the previous chapter) three approaches have been identified: the mathematical approach (in Analytical Cartography), the cognitive approach (in Cartographic Communication), and the semiotic approach (in Cartographic Language). Cartographic Visualisation uses a cognitive-semiotic approach.

We propose that these tendencies are being consolidated within the empiricist paradigm mentioned by Edney (2007). During the second half of the twentieth century, cartographic tendencies that are autonomous of each other can be recognised.

When the concepts criteria of contrast and epistemological space are applied to the tendencies belonging to the scientific-empirical paradigm, they can be recognised as paradigmatic shifts in Kuhn's terminology (1970). As shown in Fig. 7.3 of the previous chapter, there is a clear separation between the tendencies that belong to the positivism/empiricism coordinates and those in the category formed by social construction, cartographic power and the new practices, i.e. critical cartography. This recent movement is bordered by the idealism/hermeneutic and realism/post-structuralism epistemological coordinates.

Based on the aforementioned statements it is furthermore proposed that the term 'empiricist paradigm' can be substituted by 'scientific-empirical paradigm'. The name' scientific paradigm' is generic—as it is used in many sciences and disciplines, but intrinsically there exist four paradigmatic cartographic tendencies in Kuhnian terms. The second paradigm, i.e. the critical paradigm, also has a generic name because several social disciplines go under its reference frame. In cartography, however, specific map conceptions have been suggested in this critical context.

In philosophical and epistemological terms this paradigmatic differentiation is supported by Wittgenstein's philosophy. The statements of Wittgenstein's First Philosophy—the language-world relationship—are in agreement with modern or scientific cartography. The Positivist and Neo-Positivist approaches in mapping run along this line. The majority of tendencies that conceive the map as an exact reflection of reality are framed within Wittgenstein's perspective. In turn, critical cartography, in the post-modern context, represents Wittgenstein's later legacy in which language aspects are associated with the conception of pragmatic maps. Thus, the critical approach in cartography can also be interpreted in Wittengstein's terms.

These two contrasting paradigms in cartography can be compared with the positivism and historicism (or anti-positivism) stances established in geography by Capel (1983). The author grouped the diversity of geographic traditions inside these two big tendencies that have a *pendular movement*. Assuming that cartography is a rather recent scientific discipline—from the second half of the twentieth century—so far this pendulum movement has only one direction in the discipline: from positivism towards anti-positivism.

If the empiricist paradigm (or paradigmatic tendencies within the scientific-empirical perspective) has been a milestone in the development of cartography, then the transition to the next paradigmatic tendency, i.e. critical cartography, must be taken into account as a contribution to the body of knowledge of the discipline.

8.2.3 Transition from the Scientific Paradigm to the Critical Paradigm

Alan MacEachren suggested a new perspective that was quite opposite to the traditional communication paradigm. He proposed a representational view of cartography considering three dimensions: visual perception, visual cognition, and semiotics. The author identified differences between cartographic representation and visualisation, and proposed two primary-level analyses: public/social and private/perceptual-cognitive (MacEachren 1995).

From a semiotic viewpoint, MacEachren declared that syntactics and semantics belong to the private and pragmatics to the public realm. In other words, MacEachren's insight is about cartographic representation in a visualisation context with a cognitive-semiotic approach. He further deepened the topic of cartography and mapping as representations of visual communication through new theoretical and methodological tools.

On the other hand, Thomasz Zarycki suggested a semiotic approach in the study of maps. He critically analysed MacEachren's work from a (carto)semiotic perspective (Zarycki 2001a), concluding that MacEachren did not consider the importance of the pragmatic aspect by emphasizing only the syntactic and semantic dimensions. Zarycki showed remarkable differences between a *semantic map* and a *pragmatic map* which are feasible to be distinguished from an epistemological viewpoint. In other words, both map types are sufficiently different to establish a different conception of maps.

The aforementioned statement implies that the pragmatic aspect of maps belongs to a different perspective, one that is external to the scientific approach in cartography. This means that the characteristics of a pragmatic map do not correspond to a scientific approach. These maps are far from being objective, secure, neutral and value-free artefacts.

Considering other aspects, Georg Gartner established several differences in the virtual platform context between Web-mapping in Web 1.0 and the semiotics of Web 2.0. He claims that the major change is a new emphasis in the semiotic dimensions. Thus the former implies syntactic and semantic approaches, and the latter one full potential in pragmatics dimension research (Gartner 2009).

Web mapping 2.0 enables the integration of social and technical aspects into models of cartographic communication, and the process of technological change is in itself leading to an important rethinking of mapping' (Gartner 2009: 68). In other words, nowadays there exists both a *collaborative* and a *participative* nature of web mapping in Web 2.0, which was not possible with Web 1.0, allowing in virtual platform environments a change in the research focus during the communication process towards user behaviour and interests.

During the first decade of the twenty-first century, a real democratisation of cartography and mapping occurred with the big step forward of geo-technologies in the context of the World Wide Web. In this context—as shown in the chapter "Critical Cartography in the Context of Post-Modernism"—authors such as Jeremy Crampton and John Krygier point out 'mapping practices' (Crampton and Krygier 2006) and Chris Perkins refers to 'collaborative community mapping' (Perkins 2007). All these practices, which include the major participation of the user, have been consolidated from outside of the scientific-professional field. This heralds a change in focus from the cartographer/mapmaker to the user, in which users themselves can design and construct their own maps.

This democratisation of mapping implies that cartography as a practice has been 'liberated' from its academic context. Thus, cartography is no longer the patrimony of professional cartographers, and map design has escaped from the formal and strict rules that have dominated it during the scientific approach period.

Returning to Zarycki's critiques of MacEachren's approach, the lack of pragmatic depth in MacEachren's perspective is due to the fact that it is situated within the scientific-empirical paradigm of cartography. This paradigm does not consider mapping according to a social context as requested in critical social theory. Rather, the semantic map characteristics described by Zarycki belong to the syntactic and semantic dimensions. However, features of *pragmatic maps* accord to social theory, which does not share the scientific approach. Thus this new approach is framed in a post-structuralist perspective.

MacEachren's contributions can be considered as the last comprehensive theoretical insight into the scientific perspective in cartography before it underwent a strong counterweight from the critical paradigm. In fact, his opinion on the lexical approach to map representation through the pragmatic dimension is not sufficient to embed all of the implied aspects. In other words, MacEachren's stance has an epistemological difference that cannot be reconciled with the pragmatic perspective on mapping.

Consequently, when Gartner highlights potential cartographic research in web mapping 2.0 with regard to the pragmatics dimension, his call is concordant with the current context mentioned by Crampton and Krygier as well as Perkins on mapping practices and collaborative community mapping. Also, there are coincidences between Gartner and Zarycki when the latter one mentions map making as actions or when he compares the transmission of information—the traditional approach in cartography—versus the interaction between map makers and map users.

If pragmatics is considered to be the study of 'the relations of signs with their interpreters' (or the relationship between linguistic forms and their users)—according to Morris (1938) cited by Zarycki (2001b: 3099), then in this book, it is proposed that the scope of the semiotic approach clearly parallels the democratisation of cartography where the user has become a key player. Indeed, nowadays there are more and more users designing, creating, and interpreting maps. Therefore, within cartographic semiotics, i.e. syntactic—semantic—pragmatic, the pragmatic approach of map representation has become an important element in this triad.

To our minds, we clearly see a transition between the scientific-empirical paradigm and the critical one. In cartography, this stage is analysed by MacEachren's cognitive-semiotic approach of visualisation. Criticism of this approach arose due to the superficial analysis in semiotic cartography, especially of its pragmatic dimension. The emphasis that Zarycki has put on the pragmatic interpretation in mapping concerns its semiotic dimension. This perspective is closer to the critical paradigm. Similarly, in a technological context this stance is empowered by Gartner's posture on the social dimension in the distributed digital tools used in mapping and cartography.

This transition can be interpreted as a crisis or revolution in Kuhnian terms, and this great change brings a renovation in map conceptions. For example, the statements that the map as an objective, scientific representation or the "map is transparent and ideologically neutral—as in the traditional scientific cartography—is changed to conceptions such as the maps as an ideologically laden representation (Harley 1989) or the map as an historical product operating within a certain horizon of possibilities (Crampton 2003).

Harley's legacy, and his use of Foucault's power-knowledge relationships and Derrida's hermeneutic approach applied to maps, has opened discussions for new developments and insights in cartography. Thus, at the end of the twentieth century, these new map conceptions supported by information communication technologies give an account of alternative and radical visions that can be considered to be a paradigmatic shift.

8.2.4 Post-Representational Cartography: A Paradigmatic Proposal

The transition between the Scientific-Empirical Paradigm and Post-Representational Cartography, or the transition from critical paradigm to post-representational cartography, is not yet entirely clear, but attempts have been made to clarify it. In his analysis of mapping practices, Rob Kitchin compares different map ontologies and writes that the work of Denis Wood and John Fels are a transition between Brian Harley's critical cartography and that of Rob Kitchin and Martin Dodge, who are representatives of post-representational cartography (see Kitchin 2008). This transition from an ontologically secure map—taken for granted from the time of Arthur Robinson to that of Harley—to the ontologically insecure mapping was analysed in detail in the Chapter "Post-Representational Cartography".

The proposal for a post-representational cartography was first set forth by John Pickles and was then further elaborated by Rob Kitchin, Chris Perkins, and Martin Dodge (Pickles 2004; Kitchin et al. 2009). In Fig. 7.4 of the previous chapter, the tendency of post-representational cartography is located along the emergent coordinates (the ontogenesis level) within the epistemological space. This figure also indicates some authors who can be considered transitional between the constructivist approach and the emergent one like Emanuela Casti, Bruno Latour, Veronica Della Dora, James Corner, Tim Ingold, Vincent Del Casino and Stephen Hanna. Indeed, these authors have to be located in a transition zone from a critical paradigm to the post-representational stance.

When analysing the criteria of contrast mentioned earlier in this book, the cartographic tendencies from cartographic language to critical cartography have an implicit common feature: they refer to an *ontologically secure map* (cf. ontological level in Table 8.2). Therefore, representational cartography, conceived as a set of scientific and critical tendencies, receives validation in an ontological sense.

Post-representational cartography goes beyond the distinction between the empiricist paradigm and the critical paradigm with regard to the cartographic tendencies identified in this study (cartographic language, cartographic communication,

Table 8.2 Cartographic tendencies associated to paradigmatic shifts according to epistemological levels (after Azócar 2012)

Cartographic tendencies (1) (scientific level; Essentialist)	Cartographic language	Cartographic communication	Analytical cartography	Cartographic visualisation
Cartographic tendencies (2) (sociological level; Constructed)	Scientific-empirical paradigm (a)	Critical paradigm (b)	-	-
Cartographic tendencies (3) (ontological level; Emergent)	Representational paradigm (c)	Post-representational paradigm	-	-

(a) includes Cartographic tendencies (1)
(c) includes (a) (b)

analytical cartography, cartographic visualisation, and critical cartography). So, the new cartographic practices mentioned by social theory authors, e.g. counter-mapping, ethno-cartography, collaborative mapping (cf. Table 5.1) can be located outside the framework of cartographic representation. In this way, post-representational cartography considers the map to be ontologically insecure, meaning that the map is emergent and mutable depending to the context.

Post-representational cartography led to rethinking map principles, both ontologically and epistemologically. Several map concepts conceived during the post-modern period of the discipline such as maps as inscriptions or maps as unstable and complex texts (Pickles 2004), or maps as spatial practices that do work in the world, and maps as suite of cultural practices involving actions and affects (Kitchin and Dodge 2007) are examples of the rethinking about maps that has been mentioned above. All these map conceptions are opposite to the traditional ideas, i.e. those of the scientific approach (Table 5.2) exemplified by maps as objective, scientific representations, or maps being transparent and ideologically neutral, in other words: maps as truth.

8.3 Paradigms in Cartography: An Epistemological Review

8.3.1 Returning to Kuhn

Six stances in cartography can be considered as paradigmatic tendencies when the results of the application of the criteria of contrast (listed in Table 8.1) are analysed. This may imply that each stance is a normal science period in Kuhnian terminology. If the analysis is carried out at a scientific level (with essentialist features) from an epistemological perspective, then four of these tendencies will be paradigmatic. These are: cartographic language, cartographic communication, analytical cartography and, cartographic visualisation (Table 8.2). It is, however, important to note that this is an analysis at the internal level of the discipline. In other words, cartography shows four paradigms in scientific terms—or as a scientific discipline—during the second half of the twentieth century.

Thomas Kuhn highlighted the sociological character of the paradigm theory. For him, scientific activity takes place within a set of discourses. This statement can be related to Wittgenstein's term: "language games" (Richter 2004a). Thus, scientific discourse is one more language game within a myriad of games, e.g. political, religious, cultural, etc. Wittgenstein considered the social realm when referring to the different games, leading to a sociological perspective (being socially constructed).

The traditional paradigmatic tendencies in cartography could now be grouped in a context called the scientific-empirical paradigm because of the emergence of the critical approach. This critical paradigm arose from outside of the cartographic scientific community and criticised the epistemological, philosophical, and methodological basis of all the previous tendencies. In Kuhnian conceptualisation this is a critical paradigm, which by criticising the previous tendencies, leads the discipline to become Critical Cartography. Therefore, in sociological terms cartography contains two paradigms: a critical paradigm in opposition to the scientific-empirical paradigm (cf. Table 8.2).

Thus, in agreement with Edney (2007), a paradigmatic shift between the scientific-empirical and the critical stances is proposed. Edney affirms that in the history of contemporary cartography these two paradigms can be found: empiricist and critical. This implies that within the scientific community (considering scientists in social theory) there exists a transformation of the worldview as pointed out by Romaniuk and Paillalef (2010), who pose a change in the worldview not only because of changes in the modes of speaking, seeing, and knowing the world, but also a change in the modes in which problems arise, and the modes of work, rules, instruments, and techniques.

The aforementioned implies that a transition or revolutionary break between the scientific perspective and the critical one shows the following changes (cf. Table 8.1): map content as a study object, for instance, is a new manner of understanding the world through cartography and mapping. In this way, uncovering the power-knowledge practices embedded in maps—as new research aims to do—is another mode of seeing and of speaking about the world. Similarly, the use of hermeneutic and deconstructivist approaches—posed by postmodernist thinkers like Foucault and Derrida—is another mode of using rules, instruments, and techniques of work. These new modes become an alternative to the scientific-empirical perspectives. This new vision conceives the map as a social construction and as an instrument that exerts power (Harley 1989). Here, a new alternative of knowing the world is developed, using the map in a reliable way as a context-dependent device or artefact. This means that between the scientific-empirical paradigm and the critical paradigm there has been a transformation in the worldview in which different problems, lexical, ontological, gnoseological, methodological, and also practical assumptions have arisen during the last two decades, especially in Western culture.

The concept of *incommensurability* developed by Kuhn is also applicable to this new way of analysing cartography, and therefore conceptions, methods, and techniques inherent to the scientific paradigm are incompatible with critical cartographic ideas. This must be interpreted with a different logic regarding scientific cartography for a better understanding of the critical literature in cartography and mapping. Paraphrasing Kuhn: "even though practitioners of both paradigms are looking at the same data [in this case, at the *same map*], they see different worlds" (Kuhn 1970: 115).

Thus we propose that considering an ontological level of analysis, a post-representational paradigm is arising in opposition to the representational perspective (shown in Table 8.2).

Post-representational cartography arose from the emergent perspective developed by Kitchin and Dodge (2007), and Kitchin et al. (2009). In agreement with these authors we add that this stance can be interpreted as a revolutionary change that challenges traditional cartography as a representation of the world. Indeed, this new perspective goes beyond the map content as the study object. Post-representational cartography analyses the map itself; i.e. how maps emerge through cultural, social, and spatial practices framed in an ethnographical-processual approach drawn from social theory.

As in the history of physics, when Newtonian physics was replaced by Einsteinian relativistic physics in what Kuhn considers a paradigmatic shift, a new proposal that challenges the representational conception—inherent to cartography and to the map as knowledge of the world—can be applied to the cartographic discipline. The difference from previous proposals is that the post-representational paradigm targets the ontological bases of the discipline. Traditionally, these knowledge ways—or ontological bases—had only been considered under positivist and neo-positivism perspectives and were key aspects during modernity. However, these ontological bases are being questioned in a post-modern and poststructuralist context.

In this book, the paradigms in contemporary cartography have been analysed in a Kuhnian sense and different epistemological levels have been proposed. At the first level in the essentialist perspective—denominated the scientific level—four cartographic paradigms are established (Azócar 2012). This is a local or internal level within the discipline, at which all paradigmatic tendencies are different from each other. This refers to the traditional body of knowledge of the discipline in scientific terms. At the second level, denominated the sociological level, a constructivist perspective is assumed in a social context. This is an intermediate level at which a new cartographic discourse in opposition to the scientific one arises. Thus two paradigms that oppose each other are established: the critical paradigm and the scientific-empirical paradigm. Finally, the third level—called the ontological level—is a way in which an emergent perspective of knowledge is considered. This is an external level that goes beyond the scientific and sociological stances. In consequence, the critical paradigm and the scientific-empirical paradigm are merged, because both worldviews share an ontologically secure map. Thus, the post-representational paradigm arises as a new ontological proposal. As a result, two new paradigms are proposed: a *Representational Paradigm* and a *Post-Representational Paradigm*, which correspond to the modern and post-modern periods in cartography, respectively (cf. Azócar 2012). Thus, the post-representational paradigm arises as a new ontological proposal in cartography in opposition to the representational tradition.

8.3.2 On the Scope of Cartographic Paradigms

Three aspects or characteristics about paradigms can be integrated to analyse the concept of paradigm in cartography. First, the character of the science or discipline of cartography has been extensively discussed. Independent of its status, the term *disciplinary communities*, i.e. cartographic disciplinary communities, may be used instead of *scientific community*, i.e. cartographic scientific community, according to García Sierra (1999). Second, two paradigms can survive side-by-side at the same time (Hall 2006). Third, a paradigm is what is shared by members of a disciplinary community in general, of which science is just one more community (Kuhn 1970).

Additionally, a paradigm is a conceptual tool that is applicable to the sciences and disciplines across their philosophical, epistemological, sociological, historical, methodological, and practical bases. This implies that all scientific and disciplinary communities have generic paradigms, but the name or adjective given to the paradigms is related to the subjects and contents embraced by the specific field of knowledge. In this sense, there are no paradigms that are strictly unique to each discipline or science. The name or adjective of the paradigm is related to the subjects and contents embraced by a specific field of knowledge.

In this book, the analysed period of cartography's development is brief when compared to that of the physical and social sciences, which have had a more extended theoretical and practical development: it encompasses only six decades. Nevertheless, scientific cartography and mapping have their bases in the positivist period of the eighteenth century. This book covers a historic interval from modernity to post-modernity. Therefore, this time is enough to appropriately frame the discipline on the three bases of modernism and post-modernism (philosophical, epistemological and ontological).

In this way, three paradigmatic shifts in Kuhnian terms can be distinguished considering the development of cartography and mapping in an epistemological perspective: a scientific-empirical, a critical, and a post-representational paradigm (cf. Azócar 2012). The scientific-empirical paradigm of the discipline is consolidated in contemporary cartography by the following internal paradigmatic tendencies: cartographic language, cartographic communication, analytical cartography, and cartographic visualisation. Similarly, in a sociological context, the critical approach has all the features to be considered a critical paradigm opposing the scientific paradigm. Finally, in ontological terms a post-representational paradigm is proposed in opposition to the representational tradition, including the critical and scientific-empirical paradigms.

To sum up the discussion, four paradigmatic indicators can be proposed in an internal scientific context in addition to the three paradigmatic-shifts mentioned above: cartographic language, cartographic communication, analytical cartography, and cartographic visualisation. Two paradigms are defined by sociological context: scientific-empirical paradigm and critical paradigm. And two paradigms are established according to ontological terms: representational cartography and post-representational cartography. This proposal is based upon Kuhn's statements. Which paradigmatic stance will be consolidated? According to Kuhn, a paradigm is dependent on the sociological context; it is dependent on the historical force of its discourse; and it is dependent on the solidity of the disciplinary community. In other words, the consolidation of paradigms materialises through the persuasive power of the social acts. So, the consolidation of paradigms in cartography will depend on the persuasive power of the cartographic communities.

Chapter 9
Conclusions

9.1 Insights

In this book the philosophical and epistemological bases of cartography during their contemporary development were analysed. During modernity, cartography as a scientific discipline gave rise to several map conceptions under the positivism-empiricism and neo-positivism frameworks. Nevertheless, current map conceptions have also been influenced by post-modernist and poststructuralist perspectives.

From a philosophical viewpoint, the investigations showed that there is a link between Wittgenstein's philosophy and cartography. His philosophical contribution and legacy, his First Philosophy (language-world relationships) and his second one (language games), can be compared to the modernist and post-modernist cartographic development respectively. This means that within the first period, cartography followed strict rules to achieve formal statements and generalisations (i.e. semantic maps), whereas during the second period cartography and mapping went beyond academic discourse and attempts in the understanding of reality, considering different social contexts (i.e. pragmatic maps). In other words, the positivist/neo-positivist approach and the critical approach in mapping can be interpreted in Wittgenstein's terms.

This study also contended that Popper's Three Worlds Theory is linked to cartography and maps in the context of the neo-positivism approach. The construction of new realities through maps (ontological approach), and the information—transfer model of maps (cognitive approach) give account of elements belonging to Popper's World 3. For Popper, his so-called World 3 is the world of the contents of thought and the products of the human mind. Cartographically, products such as map mental images, map model and map language fall into this category (cf. Buchroithner 1997). These objects are *as real as* the objects belonging to the world of physical objects, i.e. a map device. Additionally, the distinction between phenomena-representations (P-reps) and concept-representations (C-reps) in the context of visualisation (from Ganter, cited by MacEachren 1994) also fall into this categorisation. The former ones are intended to represent the physical world (i.e. World 1) and the latter ones refer not to a thing (a physical object) but to an idea (i.e. World 3).

P. I. Azócar Fernández and M. F. Buchroithner, *Paradigms in Cartography*,
DOI: 10.1007/978-3-642-38893-4_9, © Springer-Verlag Berlin Heidelberg 2014

As the history of sciences, the development of cartography also can be analysed under the scientific revolution stance and/or normal science periods according to the paradigm concept of Thomas Kuhn. This implies that cartography—as a way of knowledge—conceives a worldview according to the historical-social context. So, cartography is integrated by a scientific or disciplinary community, in which one or two paradigms are shared by the members of this community. In sociological terms, if cartography and mapping are considered part of a scientific discourse, then they are an arbitrary way of understanding reality.

This book reviews the state-of-the-art in cartography and mapping from the second half of the twentieth century until the first decade of the twenty-first century. This review revealed that some authors and theoreticians consider trends, perspectives, and approaches as explicit paradigms, some as implicit paradigms, or simply termed them tendencies. It is proposed that the development of tendencies in cartography is framed in the philosophical and epistemological fundamentals of modernity and post-modernity, i.e. positivism and empiricism, neo-positivism (logical positivism), and post-structuralism. In this way, cartographic tendencies, paradigmatic or not, are concordant with the evolution of the scientific thought and the social context. On the one hand, traditional trends like cartographic communication, analytical cartography, and cartographic visualisation arose under the positivist and neo-positivist perspectives of modernity; on the other hand, critical cartography arose in the post-modernist context.

9.2 Achievements

In a methodological attempt, this book examines the scope of paradigms in cartography through the application of "criteria of contrast" and tendency distribution in the epistemological space. The criteria of contrast imply the analysis of cartographic tendencies according to criteria such as: study object, research aims, methods, and techniques or approaches, research results, and cartographic product results. The tendency distribution in the epistemological space (Chap. 7) considers cartographic tendencies under positivism-empiricism, realism-structuralism, and idealism-hermeneutic epistemological aspects (adapted from Sheppard 2005).

Both indicators mentioned above, criteria of contrast and tendency distribution in the epistemological space, allow us to consider the traditional cartographic trends as paradigmatic tendencies in Kuhnian terminology (Azócar 2012). According to the criteria of contrast, each tendency is different from the other. Similarly, some tendencies are grouped within the epistemological space, whereas others remain isolated. This implies that the tendencies are framed in different epistemological and philosophical perspectives. In a first approximation, the tendencies in contemporary cartography analysed by the criteria of contrast are: cartographic language, cartographic communication, analytical cartography, cartographic visualisation, critical cartography, and post-representational cartography.

A second approximation was made by analysing cartographic tendencies at three epistemological levels (adapted from Kitchin 2008): scientific (essentialist or ontic), sociological (constructed), and ontological (emergent) (Azócar 2012). First, four paradigmatic tendencies were identified at a scientific level: cartographic language (cartosemiotics), cartographic communication, analytical cartography (cartographic modelling), and cartographic visualisation. These are internal paradigms that approach cartography as a scientific discipline. Two paradigms are identified at a sociological level: a scientific-empirical paradigm, which incorporates the aforementioned paradigmatic tendencies, and a critical paradigm. According to Kuhn, the scientific activity is one more discourse within a universe of discourses. Critical cartography arises as a paradigm in opposition to the traditional scientific discourse that had been taken for granted in the discipline during the first half of the twentieth century. Finally, at an ontological level the ontological security of maps is questioned with post-representational cartography arising as a new perspective in opposition to the representational tradition. Cartography as representation is implicitly incorporated in conceptions belonging to the empiricist and critical poses. In this way, two paradigms are suggested in Kuhnian terms: the representational paradigm and the post-representational paradigm.

In summary, according to these epistemological levels (scientific, sociological, and ontological) cartography and mapping present three paradigmatic-shifts: scientific-empirical, critical and post-representational. In Kuhnian terminology, a paradigm is justified and understood by considering the conditions previous to the change. In this case, the post-representational paradigm in cartography has validity respecting the representational tradition (composed by critical and scientific perspectives). A critical paradigm in cartography is validated in opposition to the empiricist perspective. In turn, the scientific-empirical paradigm is constituted by internal and traditional cartographic tendencies (as described above).

9.3 Suggested Advancements

The suggested proposal regarding paradigmatic tendencies in cartography and mapping can be refined, especially when more extended scientific and disciplinary communities are considered. In the same way, the proposal of post-representational cartography as a paradigmatic-shift has to be advanced.

An in-depth review of the underlying thinking in cartography is suggested. This intellectual task may help to locate thinkers and theoreticians within paradigmatic tendencies. For instance, the discussion of Zarycki's critique (2001a) to MacEachren's approach to map semiotics, as materialised in this book, makes sense when both statements are considered to belong to different paradigms or to a transitional stage between two paradigmatic visions.

Kuhn's concept of paradigm can be reviewed and analysed backwards in the history of cartographic representations. The current investigations were limited to the scientific-academic period of the discipline as it only encompasses a period of time beginning in the second half of the 20th century. There exists, however, a detailed history of cartography documented by J.B. Harley and David Woodward (1987), and Matthew Edney (2007) among others. These contributions could be used as a conceptual framework for further reviews in Kuhnian terms.

They have, however, to be linked with theoretical considerations like the ones by Dietz et al. (2012) which surprisingly enough lead back to the constructs of the great theorists in cartography like Karl Peucker, Max Eckert, Kolacný or Ratajski.

Finally, the links between the development of cartography and the development of philosophy and epistemology must be highlighted through additional theoretical research. This will certainly help to strengthen the body of knowledge of the cartographic discipline!

References

Abdalla, A. (2013). Personal GIS—Ein Tool zur Verarbeitung unserer persönlichen Daten. Personal GIS—A tool to process our personal data. *Kartographische Nachrichten, 2/3*, 86–88.

Abler, R., Adams, J., & Gould, P. (1971). *Spatial organisation. The geographer's views of the world*. Englewood Cliffs: Prentice-Hall.

Amodeo, C. (2005). Late great geographers: Immanuel Kant (1724–1804). *Geographical, 77*(3), 9.

Andrieko, G., Andrieko, N., Jankowski, P., Keim, D., Kraak, M. J., MacEachren, A., et al. (2007). Visual analytics for spatial decision support. Setting the research agenda. *International Journal of Geographic Information Science,8*, 839–857.

Arnberger, E. (1970). *Die Kartographie als Wissenschaft und ihre Beziehungen zur Geographie und Geodäsie*. Grundsatzfragen der Kartographie. Vienna: Deuticke Publishers.

Azócar, P. (2006a). Teoría Cartográfica: Contextualización Científica y Formalización de la Disciplina. *Revista Semestral Humanidades y Educación,1*(2), 37–43. Departamento de Humanidades, Universidad Tecnológica Metropolitana. Santiago de Chile.

Azócar, P. (2006b). Hacia una Formalización Teórica en Cartografía: Temas y Principios de la Praxis Cartográfica. *Revista Cartográfica,82*, 75–91.

Azócar, P. (2007–2008). La Cartografía en el contexto de las ciencias: Aproximaciones teóricas de la disciplina. *Revista Geográfica de Chile Terra Australis, 51–52*, 133–140.

Azócar, P., & Buchroithner, M. F. (2009). Cartography in the context of sciences: Theoretical and technological considerations. *Proceedings 24*th *International Cartographic Conference, Santiago de Chile, November 2011*, 7 p. CD-ROM.

Azócar, P. (2012). *Paradigmatic tendencies in cartography: A synthesis of the scientific-empirical, critical and post-representational perspectives*. Doctoral thesis. Dresden University of Technology, Faculty of Environmental Sciences. Dresden, Germany. http://nbn-resolving.de/urn:nbn:de:bsz:14-qucosa-83806.

Baudrillard, J. (1981). *Simulacres et simulations*. Paris: Galilée.

Belyea, B. (1992). Images of power: Derrida, Foucault, Harley. *Cartographica,29*(2), 1–9.

Benton, S. A., Duston-Roberge, D. A., & Simard, R. (1985). Achromatic holographic stereogram of landsat multispectral scanner data. *Optical Eng*ineering, *24*(2), 242338. http://dx.doi.org/10.1117/12.7973480.

Berg, L. (1993). Between modernism and postmodernism. *Progress in Human Geography,17*(4), 490–507.

Berque, A. (2008). *La Pensée paysagère*. Paris: Archibooks.

Berque, A. (2010). *Histoire de l'habitat ideal. De l'Orient vers l'Occident*. Paris: Le Félin.

Berque, A. (2013). Die Transgression der Karten. In M. Picker, V Maleval & F. Gabaude (Eds.), *Die Zukunft der Kartographie*. Neue und nicht so neue epistemologische Krisen. (pp. 241–256). Bielefeld: transcipt Verlag.

P. I. Azócar Fernández and M. F. Buchroithner, *Paradigms in Cartography*,
DOI: 10.1007/978-3-642-38893-4, © Springer-Verlag Berlin Heidelberg 2014

Besse, J.-M. (2008). Cartographie et pensée visuelle. Réflections sur la schématisation graphique. In I. Laboulais (Ed.), *Les usages des cartes (XVIIe–XIXe siècle). Pour une approche pragmatique des productions cartographiques* (pp. 19–32). Strasbourg: Presses universitaires de Strasbourg.

Besse, J.-M. (2010). Approches spatiales dans l'histoire des sciences et des arts. *L'espace géographique,39*, 211–224.

Besse, J.-M. (2013). Tafeln, Massstäbe, Schachteln, Bäume. Zum Gebrauch einiger räumlicher Schemata in der neuzeitlichen Geographie. In M. Picker, V. Maleval & F. Gabaude (Eds.), *Die Zukunft der Kartographie. Neue und nicht so neue epistemologische Krisen* (pp. 109–133). Bielefeld: transcipt Verlag.

Board, Ch. (1967). Maps as Models. In R. J. Chorley & P. Haggett (Eds.), *Models in Geography* (pp. 671–725). London: Methuen.

Board, Ch. (1972). Cartographic Communication. *Cartographica*, 18(2):42–78.

Bollmann, J., & Koch, W. G. (Eds.). (2001, 2002). *Lexikon der Kartographie und Geomatik, 2 Vosl.* Heidelberg: Spektrum Akademischer Verlag.

Brewton, V. (2002). *Literary Theory*. Internet Encyclopedia of Philosophy IEP A Peer-Reviewed Academic Resources. http://www.iep.utm.edu/literarytheory.htm Retrieved November, 2010.

Briones, G. (1989). *Epistemología de la Investigación*. Modulo I: Métodos y técnicas avanzadas de investigación aplicada a la educación y a las ciencias sociales. Curso de Educación a Distancia. Santiago, Chile.

Bröhmer, K., Knust, C., Dickmann, F., & Buchroithner, M. F. (2012). Die Nutzung autostereoskopischer Monitore zur kartographischen Visualisierung von Diagrammen. *Kartographische Nachrichten,5*, 227–234.

Bruhm, K., Buchroithner, M. F., & Hetze, B. (2010). True-3D visualization of glacier retreat in the dachstein massif, Austria: Cross-media hard- and softcopy displays. In T. Neutens & P. Maeyer (Eds.), *Developments in 3D Geo-Information Sciences* (pp. 17–32) Springer Lecture Notes in Geoinformation and Cartography Berlin: Springer.

Brunner-Friedrich, B. & Stadler, A. (2006). Optimale Systemausrichtung eines interaktiven KIS. In J. Strobl, T. Blaschke & G. Griesebener (Eds.), *Angewandte Geoinformatik 2006. Beiträge zum 18. AGIT-Symposium*: 18–22. Heidelberg: Wichmann.

Buchroithner, M. F. (1997). Multimediale Kartographie, Data Integration und Information Fusion. In: GIS und Kartographie im multimedialen Umfeld. *Kartographische Schriften,2*, 31–35.

Buchroithner, M. F. (1998). Erstes stereoskopisches kartographisches Display. *Photogrammetrie—Fernerkundung—Geoinformation, 6,* 378–379.

Buchroithner, M. F. (1999). Möglichkeiten "echter" 3D-Visualisierungen von Geo-Daten. In *Angewandte Geographische Informationsverarbeitung XI*, (pp. 86–93) Herbert Wichmann Verlag, Heidelberg, Conference Volume AGIT Symposium Salzburg.

Buchroithner, M. F. (2000). Erste Holographische Hochgebirgskarte. In J. Aschenbrenner & H. Suida (Eds.), *1960–2000 40 Jahre glaziologische Forschung. Festschrift für Heinz SLUPETZKY zum 60. Geburtstag*. Salzburger Geographische Arbeiten, 36: 39–47.

Buchroithner, M. F. (2005). Interactive Real-Time VR Cartography. In *Proceedings of the 22nd International Cartographic Conference, 9-16 July 2005, La Coruña, Spain* (CD-ROM).

Buchroithner, M. F. (2007). Echtdreidimensionalität in der Kartographie: Gestern, heute und morgen. *Kartographische Nachrichten,57*(5), 239–248.

Buchroithner, M. F. (2008). We make you see the world three-D: Lenticular geo-displays. *Proceedings 13th International Conference on Geometry and Graphics. Dresden, August 2008*. 8 p. Dresden: ISCC.

Buchroithner, M. F. (2011). Kartographie—eine selbständige, integrative und spannende Wissenschaft. *Österreichische Zeitschrift für Vermessung und Geoinformation,1*, 24–33.

Buchroithner, M. F. (2012). Cogito ubi sum. Ein Plädoyer für gute, aktuelle Gebirgskarten (und deren Benutzung). In D. Beineke, O. Heunecke, Th. Horst & U. G. F. Kleim (Eds.), *Festschrift für Univ.-Prof. Dr.-Ing. Kurt Brunner anlässlich des Ausscheidens aus dem aktiven Dienst*. Schriftenreihe Institut für Geodäsie, *87*, 31–38. Munich: University of Armed Forces.

Buchroithner, M. F., & Azócar Fernández, P. (2011). Cartography in the context of sciences: Theoretical and technological considerations. *The Cartographic Journal,48*(1), 4–10.

Buchroithner, M. F. & Böhm, R. (1998). Cartographic relief representations with DEM filtering techniques. *ISPRS International Archives of the Photogrammetry, Remote Sensing and Spatial Information Sciences*, *32*(7), 627–632. Budapest: ISPRS.

Buchroithner, M. F., Boulos, M. N., & Robinson, L. R. (2012). Stereoscopic 3-D solutions for online maps and virtual globes. In M. F. Buchroithner (Ed.), *True-3D in Cartography—Autostereoscopic and Solid Visualisation of Geodata* (pp. 391–412)., Springer Lecture Notes in Geoinformation and Cartography Berlin: Springer.

Buchroithner, M. F. & Gartner, G. (2013). The New Face of Cartography. *GIM International. The Global Magazine for Geomatics*, 27(6): 22–27.

Buchroithner, M. F., Gründemann, T., Habermann, K. (2004a). Theoretical modelling aspects of bitemporal true-3D hardcopy production. *Honorary Volume at the Occasion of the 60th Birthday of Prof. Dr. Armin Gruen* (pp. 33–38). Zurich: ETH, Department of Geodesy and Photogrammetry.

Buchroithner, M. F., Gründemann, T., Habermann, K., König, K. & Walder, O. (2004b). Theoretische Aspekte der Modellierung von Echt-3D-Darstellungen. In *Theorie 2003. Vorträge der 8. Dresdner Sommerschule für Kartographie am 25./26.* (pp. 62–74) *September 2003*, Kartographische Bausteine, Institute for Cartography, Dresden University of Technology, Germany.

Buchroithner, M. F., Gruendemann, T., Kirk, R. L., & Habermann, K. (2005a). Three in one: Multiscale hardcopy depiction of the mars surface in true-3D. *Photogrammetric Engineering & Remote Sensing,71*(10), 1105–1108.

Buchroithner, M. F., Gruendemann, T., Habermann, K., Neukum, G. & the HRSC Co-Investigator Team (2005b). Multitemporale und großformatige Echt-3D-Hartkopien der Marsoberfläche. *Photogrammetrie—Fernerkundung—Geoinformation*, *5*, 429–434.

Buchroithner, M. F., Habermann, K. & Gruendemann, T. (2005c). Modeling of Three-Dimensional Geodata Sets for True-3D Lenticular Foil Displays. *Photogrammetrie—Fernerkundung—Geoinformation*, *1*, 47–56.

Buchroithner, M. F., & Habermann, K. (2010). The art of autostereoscopic relief representation in cartography. In G. Gartner & F. Ortag (Eds.), *Cartography in Central and Eastern Europe*., Lecture Notes in Geoinformation and Cartography Berlin: Springer.

Buchroithner, M. F. & Knust, C. (2013a). True-3D in Cartography—Current Hard- and Softcopy Developments. In A. Moore & I. Drecki (Eds.), *Geospatial Visualisation*. Springer Lecture Notes in Geoinformation and Cartography. (pp. 41–65). Berlin: Springer.

Buchroithner, M. F. & Knust. C. (2013b). Principles and Terminology of True-3D Geovisualisation.

Buchroithner, M. F. & Knust. C. (2013c). The Third Dimension in Cartography: Recent and Future Developments. Die dritte Dimension in der Kartographie: rezente und zukünftige Entwicklungen. *Kartographische Nachrichten*, 63, Special ICC 2013 Issue, 127–132.

Buchroithner, M. F., & Kostka, R. (Eds.) (1994). *TADAT-Endbericht. Vol. I and II*. Totally 747 pp. Graz: Techn. Univ.

Buchroithner, M. F., & Kostka, R. (1996). Hochgebirge, Kartographie und kosmische Fernerkundung. Einige grundsätzliche Bemerkungen. *Kartographische Nachrichten,1996*(4), 140–148.

Buchroithner, M. F. & Kostka, R. (1997). Conceptual considerations on high-mountain cartography and spaceborne remote sensing. In *Proceeding 18th ICA/ACI International Cartographic Conference, Stockholm*, Vol. 4, pp. 2056–2059.

Buchroithner, M. F., & Schenkel, R. (1999). 3D mapping with holography. *GIM International,13*(8), 32–35.

Buchroithner, M. F. & Schenkel, R. (2001). Kartographische Echt-3D-Visualisierung. Eine Ergänzung zu "flachen Karten". In Th. Wintges & R. Buzin (Eds.), *Proceedings Deutscher Kartographentag* (pp. 55–64) Berchtesgaden: Wichmann.

Buchroithner, M. F., Schenkel, R. & Kirschenbauer, S. (2000). 3D display techniques for cartographic purposes: Semiotic aspects. In *ISPRS International Archives of Photogrammetry and Remote Sensing*, *33*: 99–106. Amsterdam: ISPRS.

Bunge, M. (1998). *Philosophy of science: From problem to theory* (Vol. 1). New Brunswick: Transaction Publishers.

Capel, H. (1983). Positivismo y Anti-positivismo en la Ciencia Geográfica: El Ejemplo de la Geomorfología. *Geo-Crítica. Cuadernos Críticos de Geografía Humana*, Año VIII, N° 43. Retrieved June, 2010, from www.ub.edu/geocrit/geo43.htm.

Capel, H. (1998). La Geografía en el Siglo XXI. *Scripta Nova Revista Electrónica de Geografía y Ciencias Sociales* N° 19, Universidad de Barcelona. España.

Carey, R. (2008). *Russell's Metaphysics*. Internet Encyclopedia of Philosophy IEP, A Peer-Reviewed Academic Resource. Retrieved November, 2010, from http://www.iep.utm.edu/r/russell.htm.

Carnap, R. (1931). Die Überwindung der Metaphysik durch logische Analyse der Sprache. *Erkenntnis,2*(1), 219–241.

Cartwright, W., Gartner, G., & Lehn, A. (2009). Maps and mapping in the eyes of artists and cartographers—Experiences from the international symposium on cartography and art. In W. Cartwright, G. Gartner, & A. Lehn (Eds.), *Cartography and Art.*, Lecture Notes in Geoinformation and Cartography Berlin: Springer.

Cassettari, S., Fagg, A., & Visvalingam, M. (1992). Cartography and geographical information systems. *The Cartographic Journal,29*(1), 51–54.

Casti, E. (2005). Toward a theory of interpretation: Cartographic semiosis. *Cartographica,40*(3), 1–16.

Cauvin, C., Escobar, F., & Serradj, A. (2010). *Thematic Cartography and Transformations* (Vol. 1). London: ISTE Ltd.

Chorley, R., & Hagget, P. (1967). *Models in Geography*. London: Methuen.

Clack, B. (1999). *An Introduction Wittgenstein's Philosophy of Religion*. Edinburgh: University Press.

Clarke, K., & Cloud, J. (2000). On the origins of analytical cartography. *Cartography and Geographic Information Science,27*(3), 195–204.

Conley, T. (1996). *The Self-Made Map. Cartographic Writings in the Early Modern France*. Minneapolis: University of Minnesota Press.

Corner, J. (1999). The Agency of mapping: speculation, critique and invention. In D. Cosgrove (Ed.), *Mapping*. London: Reaktion Books.

Cosgrove, D. (1999). Introduction mapping meaning. In D. Cosgrove (Ed.), *Mappings* (pp. 1–23). London: Reaction Books.

Cosgrove, D. (2005). Mapping/cartography. In D. Atkinson et al. (Eds.), *Cultural Geography. A Critical Dictionary of Key Concepts* (pp. 27–33). London: Tauris.

Cosgrove, D. (2007). Epistemology, geography, and cartography: Matthew Edney on Brian Harley's cartographic theories. *Annals of the Association of American Geographers,97*(1), 202–209.

Crampton, J. (2001). Maps as social constructions: Power, communication and visualization. *Progress in Human Geography,25*(2), 235–252.

Crampton, J. (2002). Thinking philosophically in cartography: Toward a critical politics of mapping. *Cartographic Perspectives,41*, 4–21.

Crampton, J. (2003). *The Political Mapping of the Cyberspace*. Edinburgh: Edinburgh University Press.

Crampton, J. (2010). *Mapping: A Critical Introduction to Cartography and GIS*. Oxford: Wiley.

Crampton, J., & Elden, S. (Eds.). (2007). *Space, knowledge and power: Foucault and geography*. Hampshire: Ashgate Publishing Limited.

Crampton, J., & Krygier, J. (2006). An introduction to critical cartography. *ACME: An International E-Journal for Critical Geographies*, *4*(1), 11–33.

Cybercartography (2008). Geomatics and cartographic research centre GCRC. Otawa: Department of Geography and Environmental Studies, Carleton University. Retrieved October, 2010, from http://www.gcrc.carleton.ca/confluence/.../Cybercartography.

Dalkiran, H. P., Özağac, S., & Büyükbayrak (2009). *Proceedings of the 24th International Cartographic Conference 2009, Santiago de Chile*, 16 p. http://icaci.org/files/documents/ICC_proceedings/ICC2009/html/nonref/1_3.pdf.

Dalkiran, H. P., Özağac, S. & Büyükbayrak (2012). Holographic Relief map production by using topographic line maps (Digital Carto-Holograms). In M. F. Buchroithner (Ed.) *True-3D Cartography: Autostereoscopic and Solid Visualisation of Geodata*. Lecture Notes in Geoinformation and Cartography (pp. 367–388) Berlin: Springer.

Daston, L., & Galison, P. (2007). *Objectivity*. New York: Zone Books.

Del Casino Jr., V. & Hanna, S. (2006). Beyond the "Binaries": A methodological intervention for interrogating maps as representational practices. *ACME: An International E-Journal for Critical Geographies*, *4*(1), 34–56.

Della Dora, V. (2009). Performative atlases: Memory, materiality, and (co-) authorship. *Cartographica,45*(4), 240–255.

Derrida, J. (1976). *Of Grammatology* (G. C. Spivak, Trans.). Baltimore: John Hopkins University Press.

DiBiase, D. (1990). Visualization in the earth sciences. *Earth and Mineral Sciences*, *59*(2). Pennsylvania State University. Retrieved February, 2012, from http://www.geovista.psu.edu/publications/others/dibiase90/swoopy.html.

DiBiase, D., MacEachren, A., Krygier, J., & Reeves, C. (1992). Animation and the role of map design in scientific visualization. *Cartography and Geographic Information Systems,19*(4), 201–214, 265–266.

Dickmann, F., & Kestermann, E. (2011). Der Einfluss von Stadtplänen und Fahrzeug-Navigationssystemen auf die Entwicklung mentaler Raummodelle. *Kartographische Nachrichten,61*(4), 183–190.

Dickmann, F., Bröhmer, K., Buchroithner, M. F. & Knust, C. (2012). Möglichkeiten und Grenzen lentikularer Mehrbildmodelle im Vermittlungsprozess raumbezogener Informationen. In N. Diekmann-Boubaker, F. Dickmann (Hrsg.). Innovatives Lernen mit kartographischen Medien. *Kartographische Schriften*, *15*: 129–144. Bonn: Kirschbaum Verlag.

Dietz, R., Lechthaler, M. & Mang, R. (2012). Geostrategie und Kartenwissenschaft. *Österreichische Militärische Zeitschrift*, 6, 662–671.

Dodge, M., Kitchin, R., & Perkins C. (2011). Introductory essay: Conceptualising mapping. In M. Dodge, R. Kitchin & C. Perkins (Eds.) *The Map Reader: Theories of Mapping Practice and Cartographic Representation* (pp. 1–7). Chichester: Wiley.

Dommann, M. & Meier, M. (1999). Wissenschaft, die Bilder schafft. In: M. Dommann & M. Meiner (Eds.) *Wissenschaft, die Bilder schafft.Science en images*. Traverse. Zeitschrift für Geschichte—Revue d'histoire, 18, 1999, 15.

Dünne, J. (2011). *Die kartographische Imagination. Erinnern, Erzählen und Fingieren in der Frühen Neuzeit*. Munich: Fink.

Dünne, J. (2013): Die Unheimlichkeit des Mapping. In M. Picker, V. Maleval & F. Gabaude (Eds), *Die Zukunft der Kartographie. Neue und nicht so neue epistemologische Krisen*. (pp. 211–240). Bielefeld: transcipt Verlag.

Dym, J., & Offen, K. (2011). *Mapping Latin America. A Cartographic Reader*. Chicago: University of Chicago Press.

Eco, U. (1975). *Trattato di semiotica generale*. Milano: Bompiani.

Eco, U. (1985). *Einführung in die Semiotik*. München: Wilhelm Fink.

Eckert, M. (1907). Die Kartographie als Wissenschaft. *Zeitschrft der Gesellschaft für Erdkunde zu Berlin*, pp. 539–555.

Eckert, M. (1921, 1925). *Die Kartenwissenschaft. Forschungen und Grundlagen zu einer Kartographie als Wissenschaft*. Berlin u. Leipzig, W. de Gruyter.

Eckert, M. (1977). On the nature of maps and map logic. *The Nature of Cartographic Communication, Cartographia*, Monograph. N° 19. (W. Joerg, Trans.).

Edney, M. (1993). Cartography without progress: Reinterpreting the nature and historical development of mapmaking. *Cartographica,30*(2/3), 54–68.

Edney, M. H. (1996). Theory and the History of Cartography. *Imago Mundi. The International Journal for the History of Cartography*, *48*(1), 185–191.

Edney, M. (2007). Recent trends in the history of cartography: A selective annotated bibliography to the english-language literature. *Coordinates, Series B*, N° 6. Retrieved February, 2010, from http://purl.oclc.org/coodinates/b6.htm.

El Nabbout, Kh. (2007). Geo-visualization tools for participatory urban planning. The case of Tripoli, Lebanon. PhD Thesis, Institute for Cartography, University of Technology, Dresden, Germany.

Fabrikant, S., & Lobben, A. (2009). Introduction: Cognitive issues in geographic information visualization. *Cartographica,44*(3), 139–143.

Farinelli, F. (1996). Von der Natur der Moderne. Eine Kritik der Kartographischen Vernunft. In D. Reichert (Ed.), *Räumliches Denken* (pp. 267–301). Zurich: vdf Publ.

Farinelli, F. (2009). *De la raison cartographique*. Paris: CTHS.

Farinelli, F. (2013). Am Anfang war die Karte. In M. Picker, V. Maleval & F. Gabaude (Eds). *Die Zukunft der Kartographie. Neue und nicht so neue epistemologische Krisen.* (pp. 257–277). Bielefeld: transcipt Verlag.

Field, K. (2009). Art in c'art'ography. Art & cartography special issue, November 2009. *The Cartographic Journal*, *46*(4), Editorial Preface.

Field, K. (2013). Editorial. Cartomyopic musings. *The Cartographic Journal,50*(1), 1–7.

Foucault, M. (1969). *L' Archeologie du Savoir*. Paris: Gallimard.

Frascolla, P. (2007). *Undertanding Wittgenstein's Tractatus*. New York: Routledge.

Freitag, U. (1980). Can communication theory form the basis of a general theory of cartography? In *Nachrichten aus dem Kartem—und Vermessungswesen* (pp. 17–35) R. II, Frankfurt a.M.

Freitag, U. (1985). *Theories in cartography*. Notes for lectures at the Wuhan Technical University of Surveying and Mapping in October 1985. Berlin (reproduced as manuscript).

Freitag, U. (1991). Zur Theorie der Kartographie. Grundlagen und Tendenzen ihrer Entwicklung im deutschen Sprachraum in den letzten 40 Jahren. *Kartographische Nachrichten,41*, 42–50.

Freitag, U. (1992a). Semiotik und Kartographie. In U. Freitag (Hrsg.) *Kartographische Konzeptionen. Beiträge zur theoretischen und praktischen Kartographie 1961–1991*, (pp. 1–14). Berlin, Selbstverlag Fachbereich Geowissenschaften, Freie Universität Berlin.

Freitag, U. (1992b). Grundlagen, Aufbau und zukünftige Aufgaben der kartographischen Wissenschaft. In U. Freitag (Hrsg.) *Kartographische Konzeptionen. Beiträge zur theoretischen und praktischen Kartographie 1961–1991*, (pp. 41–56). Berlin, Selbstverlag Fachbereich Geowissenschaften, Freie Universität Berlin.

Freitag, U. (2001). Die Entwicklung der Theorie der Kartographie. In W.G. Koch (Eds.), *Theorie 2000* (pp. 4–14). Kartographische Bausteine (Vol. 19), Dresden: Technische Universität .

Friedman, J. (1989). The dialectic of reason. *International Journal of Urban and Regional Research,13*(2), 217–236.

García Sierra, P. (1999). *Diccionario Filosófico*. 174 Biblioteca Filosofía en Español. Retrieved December, 2010, from http://www.filosofia.org/filomat.

Gartner, G. (2009). Web Mapping 2.0. In M. Dodge, R. Kitchin & C. Perkins (Eds.), *Rethinking Maps: New Frontiers in Cartographic Theory* (pp. 67–82) New York: Routledge Studies in Human Geography.

Gartner, G. (2013). Location-Based Services and the Role of Modern Cartography. Location-based Services und die Rolle der modernen Kartographie. *Kartographische Nachrichten*, 63, Special ICC 2013 Issue, 169–174.

Gattei, S. (2009). *Karl Popper's Philosophy of Science: Rationality without Foundation*. New York: Routledge.

Gebaude, F. & Maleval, V. (2013). Mapping als Bildrhetorik. In M. Picker, V. Maleval & F. Gabaude (Eds.), *Die Zukunft der Kartographie. Neue und nicht so neue epistemologische Krisen* (pp. 135–157). Bielefeld: transcipt Verlag.

Geomatics and Cartographic Research Centre GCRC. (2008). Department of Geography and Environmental Studies, Carleton University, Otawa. Retrieved October, 2010, from http://www.gcrc.carleton.ca/confluence/…/Cybercartography.

Giddens, A. (1991). Structuration theory: past, present and future. In Ch G A Bryant & D. Jary (Eds.), *Giddens' in Theory of Structuration. A Critical Appreciation*. London: Routledge.

Giddens, A. (1993). *The Constitution of Society. Outline of the Theory of Structuration*. Cambridge: Polity Press.

Glasze, G. (2009). Kritische Kartographie. *Geographische Zeitschrift,97*(4), 181–191.

Glauser, J., & Kiening, Ch. (Eds.). (2007). *Text—Bild—Karte. Kartographien der Vormoderne*. Freiburg: Rombach.

Gombrich, E. H. (1984). *Bild und Auge. Neue Studien zur Psychologie der bildlichen Darstellung*. Stuttgart: Klett-Gotha.

Gómez Mendoza, J., Muñoz, J., & Ortega Cantero, N. (1988). *El Pensamiento Geográfico: Estudio Interpretativo y Antología de Textos (de Humboldt a las Tendencias Radicales)*. Madrid: Alianza Editorial.

Goodman, N. (1969). *Languages of Art. An Approach to a Theory of Symbols*. London: Oxford University Press.

Gorton, W. (2010). *The Philosophy of Social Science*. Internet Encyclopedia of Philosophy IEP, A Peer-Reviewed Academic Resource. Retrieved February, 2011, from http://www.iep.utm.edu/soc-sci/.

Gould, P., & White, R. (1974). *Mental Maps*. Harmondsworth: Penguin Books.

Grelot, J. (1994). Cartography and GIS age. *The Cartographic Journal,31*(1), 56–60.

Gründemann, T. (2004). Herstellung einer Echt-3D-Hochgebirgskarte auf der Basis des Lentikularverfahrens. Engin. Diploma Thesis, Institute for Cartography. University of Technology, Dresden.

Grünreich, D. (1992). Welche Rolle spielt die Kartographie beim Aufbau und Einsatz von Geo-Infromationssystemen? *Kartographische Nachrichten,42*(1), 1–6.

Hake, G., Grünreich, D., & Meng, L. (2002). *Kartographie: Visualisierung raum-zeitlicher Informationen* (8th ed.). Berlin: De Gruyter.

Haklay, M., Singleton, A., & Parker, C. (2008). Web mapping 2.0: The Neogeography of the GeoWeb. *Geography Compass,2*(6), 2011–2039.

Hall, W. (2006). Application holy wars or a new reformation? A fugue of the theory of knowledge.
http://www.orgs-evolutionknowledge.net/Index/ApplicHolyWars/FullProjectToNow/ Accessed December 2010.

Hallisey, E. (2005). Cartographic visualization: An assessment and epistemological review. *The Professional Geographer,57*(3), 350–364.

Hantzsch, V. (1902). Berghaus, Hermann. In: *Allgemeine Deutsche Biographie* (ADB). Band 46, Ducker & Humblot, Leipzig, (pp. 379–381).

Harley, J. B. (1987). Preface. In J. B. Harley & D. Woodward (Eds.), *The History of Cartography*, (Vol. 1, XV–XXI). Chicago: University of Chicago Press.

Harley, J. B. & Woodward, D. (1987). *The History of Cartography. Volume 1. Cartography in Prehistoriyc, Ancient, and Medieval Europe and the Mediterranean*. Chicago: University of Chicago Press.

Harley, J. B. & Woodward, D. (Eds.) (1987). *The History of Cartography*, 6 volumes, 3 so far published in 2013. Chicago: University of Chicago Press.

Harley, J. B. (1988a). Maps, knowledge, and power. In D. Cosgrove & S. Daniels (Eds.), *The Iconography of Landscape: Essays on the Symbolic Representation, Design and Use of Past Environments* (pp. 277–312). Cambridge: Cambridge University Press.

Harley, J. B. (1988b). Silences and secrecy: The hidden agenda of state cartography in early modern Europe. *Imago Mundi,40*, 111–130.

Harley, J. B. (1989). Deconstructing the map. *Cartographica,26*, 1–20.

Harley, J. B. (1990). Cartography ethics and social theory. *Cartographica,27*(2), 1–23.

Harley, J. B. (1994). New England cartography and the native Americans. In E. Baker, E. Churchill, R. D'Abate, K. Jones, V. Konrad, & H. Prins (Eds.), *Beginnings: Exploration, Culture, and Cartography in the Land of Nurumbega* (pp. 287–313). Lincoln: University of Nebraska Press.

Harley, J. B. (1997). Power and Legitimation in the English geographical atlases of the eighteenth century. In J. Wolter & R. Grimm (Eds.), *Images of the World: The Atlas through History* (pp. 161–204). New York: McGraw-Hill.

Harley, J. B. (2001). *The New Nature of Maps: Essays in the History of Cartography.* In P. Laxton (Ed.), Baltimore: Johns Hopkins University Press.

Hartshorne, R. (1958). El Concepto de Geografía como ciencia del espacio: de Kant y Humboldt a Hettner. *Documents D'analisis Geográfica* 18, 1991, 31–54 Retrieved March, 2012, from http://ddd.uab.es/pub/dag/02121573p31.pdf.

Harvey, D. (1969). *Explanation in Geography*. London: Eduard Arnold.

Hebert, L. (2002). Do Maps Create or Represent Reality? Retrieved September, 2010, from http://geography.about.com/library/misc/ucmaps.htm.

Hernández, H. (1982). La Geografía: Estado Actual y Perspectivas. *Informaciones Geográficas,29*, 3–20.

Hessen, J. (1976). *Teoría del Conocimiento*. 14ª Edición. Madrid: Espasa Calpe.

Hruby, F. (2006a). Semiotische Begründbarkeit kartographischer Signaturen—Prolegomena zum Stellenwert räumlicher Dreidimensionalität in der Kartographie am Beispiel der Signatur auf der Grundlage der Semiotik von Charles S. Pierce. Diploma Thesis, Univ. Vienna, Institute for Geography and Regional Research.

Hruby, F. (2006). *Semiotische Begründbarkeit kartographischer Signaturen*. München: Grin Publishers.

Hruby, F. & Guerrero, R.M. (2008). Kartographie im Spannungsfeld expliziter und impliziter Forschung. *Meta-carto-semiotics*—[Electronic] *Journal for Theoretical Cartography, 1*, 1–13.

Hruby, F. (2011). *Prelegomena zum Potential sphärischer Displays für die Visualisierung globaler Phänomene*. Ph.D. Thesis, 171 p. June 2011, University of Vienna, Austria.

Hruby, F. & Wolodtschenko, A. (2010). 3 Jahre/Bände meta-carto-semiotics—eine Zwischenbilanz der Herausgeber. *meta-carto-semiotics,3*, 1–9. www://meta-carto-semiotics.org.

Huffman, N. H. (1996). You can't get here there: Reconstructing the relevancy of design in postmodernism. In C. Wood & P. Keller (Eds.), *Cartographic Design: Theoretical and Practical Perspectives* (pp. 35–53). New York: Wiley.

Hurni, L., & Sell, G. (2009). Cartography and architecture: Interplay between reality and fiction. *The Cartographic Journal,46*(4), 323–332.

Husserl, E. (1996). *Die Krisis der europäischen Wissenschaften und die transzendentale Phänomenologie. Eine Einleitung in die phänomenologische Philosophie*. Hamburg: Meiner.

Ingold, T. (2000). *The Perception of the Environment: Essays in Livelihood, Dwelling and Skill*. London: Routledge.

Jackson, P. (1989). *Maps of Meaning*. London: Unwin Hyman.

Jacob, Ch. (2005). *The Souvereign Map. Theoretical Approaches in Cartography throughout History*. Chicago: University of Chicago Press.

Jacob, Ch. (1992). *L'empire des cartes: Appoche theorique de la cartographie a travers l'historie*. Paris: Albin Michel.

Jameson, F. (1990). *Postmodernism, or, The Cultural Logic of Late Capitalism*. Durham: Duke University Press.

Kahan, J. (1997). Historicism. *Renaissance Quarterly,50*(4), 1202–1202.

Kainz, W., Kriz, K., Pucher, A. & Riedl, A. (2013). On-Line Services for Visualisation and Navigation. Online-Dienste für Visualisierung und Navigation. *Kartographische Nachrichten*, 63, Special ICC 2013 Issue, 160–164.

Kanakubo, T. (1990). The development of contemporary theoretical cartography. *The Science Reports of the Tohoku University* (pp. 1–19). 7th Series (Geography) Vol. 40, N° 1. June 1990.

Kanakubo, T., Pravda, J., Rimbert, S., Freitag, U., Török, Z., Morita, T., & Board, C. (1993). The selected main theoretical issues facing cartography. *Proceedings Volume 1 of the 16th International Cartographic Conference, Cologne 3–9 May 1993* (pp. 17–25). Retrieved March, 2012, from http://icaci.org/files/documents/ICC_proceedings/ICC1993/icc1993_volume1_part1.pdf.

Kavouras, M., & Kokla, M. (2008). *Theories of Geographic Concept: Ontological Approaches to Semantic Integration*. Athens: Taylor & Francis Group.

Kelnhofer, F. (1993). Kartographie als Grundlage von GIS. ÖCG – Kommunikativ, 1993, 6: without page reference.

Kelnhofer, F. (1994). Cartographic information systems and digital map production. In: *Environment and Quality of Life in Central Europe*. Prague: International Geoscientific Union (IGU). CD-ROM.

Kelnhofer, F. (1995). *Kartographische Infrormationssysteme—Ende des Printmediums Karte?16*, 127–141. Institutsmitteilungen Institut für Geodäsie, Innsbruck: Universität Innsbruck.

Kelnhofer, F. (1996). Geographisch und/oder Krtographische Informationssysteme. In: L. Hurni (Ed.), *Kartographie im Umbruch—neue Herausforderungen, neue Technologien. Beiträge zum Kartographiekongress Interlaken 96*. Kartographische Publikationsreihe, 14: 9–26. Wabern, Switzerland: Schweizerische Gesellschaft für Kartographie.

Kelnhofer, F., & Lechthaler, M. (1994). DIGMAP—Digitales kartographisches Infromations- und Originalisierungssystem. *Kartographische Nachrichten,44*(6), 219–227.

Kent, A., & Vujakovic, P. (2011). Cartographic language: Towards a new paradigms for understanding stylistic diversity in topographic maps. *The Cartographic Journal,48*(1), 21–40.

Kirschenbauer, S. & Buchroithner, M.F. (1999). "Real" 3D Technologies for relief depiction. *Proceedings of the 19th ICA/ACI International Cartographic Conference, August 1999, Ottawa, Canada*, 7 p. (CD-ROM).

Kishimoto, H. (1975). "Mental Maps" und Kartographie. *Vermessung, Photogrammetrie, Kulturtechnik,73*(1), 62–64.

Kitchin, R. (2008). The practices of mapping. *Cartographica,43*(3), 211–215.

Kitchin, R., & Dodge, M. (2007). Rethinking maps. *Progress in Human Geography,31*(3), 331–344.

Kitchin, R., Perkins, C., & Dodge, M. (2009). Thinking about maps. In M. Dodge, R. Kitchin, & C. Perkins (Eds.), *Rethinking Maps: New Frontiers in Cartographic Theory* (pp. 1–25). London: Routledge.

Koch, T. (2008). Books comment: The natures of maps: Cartographic construction of the natural world (Denis Wood and John Fels). *Cartographic Perspective,59*, 48–51.

Koch, W. G. (1995). Concepts for the teaching of theoretical cartography. In *Cartography Crossing Borders* (pp. 1855–1862) Edited by Institut Cartogràphic de Catalunya, 17th International Cartographic Conference, Proceedings 2, Barcelona.

Koch, W. G. (2003). Cartographic theories. In A. Wolodtschenko (Ed.), *The Selected Problems of Theoretical Cartography 2002* (pp. 24–33). Dresden: Institute for Cartography, Dresden University of Technology.

Koch, W. G. (2004). Theorie und Methode in der heutigen Kartographie. In: Koch, W. G. (Ed.), *Theorie 2003. Kartographische Bausteine*, *26*, 4–19, Institute for Cartography, Dresden University of Technology, Dresden.

Koch, W. G. (2010). Eckert (seit 1934 Eckert-Greifendorff, Friedrich Eduard Max. *Sächsische Biografie*, Ed. Institut für Sächsische Geschichte und Volkskunde e.V., bearb. Von Martina Schattkowsky. Retrieved March, 2012, fromhttp://www.isgv.de/saebi/.

Kolácný, A. (1969). Cartographic information: A fundamental concept and term in modern cartography. *The Cartographic Journal,6*(1), 47–49.

Kozieł, Z. (1997). Concerning the need for development of the geomatic research method. *Geodezja i Kartografia*, XLVI, 3, 217–224.

Kozieł, Z. (2001). Geokompozycje jako ujęcia wideograficzne. *Kognitywistyka I Media w Edukacji,7*(1), 13–24.

Kozieł, Z. (2005). Geocompositions as videographic representations. *Proceedings of WMSCI—9th World Multi-Conference on Systematics, Cybernetics and Informatics, Orlando, Florida, 10–13 July,2005*, Vol. V: 360–362.

Kozieł, Z. (2006). The Syncretism of geocomposition in application to multimedia presentation of the forms of the earth surface. *Proceedings of WMSCI—10th World Multi-Conference on Systematics, Cybernetics and Informatics, Orlando, Florida, 16–19 July, 2006,* Vol. 5, 256–259.

Kozieł, Z. (2013). Istota geokompozycji jako autonomocznego bytu semiotycznego. In. Medyńska-Gulij, B. & Kubiak, J. (Ed.). *Pragmatyka w kartografii*. Prace i Sudia Kartograficzne *4*, 81–87. Poznań, Poland.

König, V. (2013). Kartographische Kommunikation, räumliche Organisation und ihre Darstellung. In M. Picker, V. Maleval & F. Gabaude (Eds.), *Die Zukunft der Kartographie. Neue und nicht so neue epistemologische Krisen* (pp. 159–175). Bielefeld: Transcipt Verlag.

Kuhn, T. (1962). *The Structure of Scientific Revolutions.* Chicago: Chicago University Press.

Kuhn, T. (1970). *The Structure of Scientific Revolutions* (2nd ed.). Chicago: Chicago University Press.

Kraak, M.-J., & Ormeling, F. (1996). *Cartography. Visualization of Geospatial Data*. Harlow: Longman.

Kraak, M.-J., & Ormeling, F. (2003). *Cartography: Visualization of Geospatial Data* (2nd ed.). Essex: Pearson Education Limited.

Kraak, M.-J. (2008). From geovisualization toward geovisual analytics. *The Cartography Journal,45*(3), 163–164.

Kraak, M.-J., & Ormeling, F. (2010). *Cartography: Visualisation of Spatial Data* (3rd ed.). New York: Guilford Press.

Krygier, J. (1995). Cartography as an art and a science. *The Cartographic Journal,30*, 3–10.

Krygier, J., & Wood, D. (2005). *Making Maps: A Visual Guide to Map Design for GIS*. New York: The Guildford Press.

Krugier, J., & Wood, D. (2009). Ce n'est pas le Monde (This is not the World). In M. Dodge, R. Kitchin, & C. Perkins (Eds.), *Rethinking Maps: New Frontiers in Cartographic Theory* (pp. 189–219). London: Routledge.

Latour, B. (1987). *Science in Action: How to Follow Scientists and Engineers through Society. Milton Keynes*. Milton Keynes: Open University Press.

Latour, B. (1990). Drawing things together. In M. Lynch & S. Woolgar (Eds.), *Representation in Scientific Practice* (pp. 131–143). Chicago: University of Chicago Press.

Latour, B. (1999). *Pandora's Hope: Essays on the Reality of Science Studies*. Cambridge: Harvard University Press.

Lechthaler, M. (2010). Interactive and multimedia atlas information system as a cartographic geo-communication platform. In G. Gartner & F. Ortag (Eds.), *Cartography in Central and Eastern Europe* (pp. 384–402). Lecture Notes in Geoinformation and Cartography Berlin: Springer.

Lee, J. (1995). Map design and GIS—A survey of map usage amongst GIS users. *The Cartographic Journal,32*(1), 33–44.

Leonov, A. (2001). Computer-made composing of geo-images. *Proceedings 20th International Cartographic Conference of ICA. Beijing, August 2001.* 1 p. CD-ROM. http://icaci.org/files/documents/ICC_proceedings/ICC2001/icc2001/file/f24031.pdf.

Liehmann, St. (2003). "Der Dresdener Echt-3D-Atlas" —Konzeption und prototypische Erstellung einer geowissenschaftlich orientierten Präsentation für das "Dresdner 3D-Display". Engin. Diploma Thesis, Institute for Cartography. University of Technology, Dresden.

Liu, S., & Palen, L. (2010). The new cartographers: Crisis map mashups and the emergence of neogeographic practices. *Cartographic and Geographic Information Science,37*(1), 69–90.

Lois, C. (2009). Imagen Cartográfica e Imaginarios Geográficos: Los Lugares y las Formas de los Mapas en Nuestra Cultural Visual. *Scripta Nova Revista Electrónica de Geografía y Ciencias Sociales* Vol. XIII, pp. 281-309. Nueva serie de Geo Crítica. Cuadernos Críticos de Geografía Humana.

Lynch, M. & Woolgar, S. (1990). *Representation in Scientific Practice.* Cambridge: MIT Press.

MacEachren, A. (1994). Visualization in modern cartography. In A. MacEachren &. D. R. Fraser Taylor (Ed.), *Visualization in Modern Cartography*, Oxford: Pergamon.

MacEachren, A. (1995). *How Maps Work: Representation, Visualization and Design.* London: Guilford.

MacEachren, A., & Kraak, M.-J. (2001). Research challenges in geovisualization. *Cartographic and Geographic Information Science,28*, 3–12.

Martínez, E., & Reyes, M. d. C. (1995). Cybercartography and Society. In D. R. Fraser Taylor (Ed.), *Cybercartography: Theory and Practice* (pp. 99–122). Amsterdam: Elsevier.

Masterson, M. (1970). The nature of a paradigm. In I. Lakatos & A. Musgrave (Eds.), *Criticism and the Growth of Knowledge* (pp. 58–89). Cambridge: Cambridge University Press.

Mejía Soto, E. (2004). Aproximación al Concepto de Objetividad en Karl Popper. Retrieved December, 2010, from http://www.gestionpolis.com.

Moellering, H. (2000). The scope and conceptual content of analytical cartography. *Cartography and Geographic Information Science,27*(3), 205–223.

Moellering, H. (2001a). Analytical cartography: past, present and future. *Proceedings, 4th Conference on Geographic Information Science, Brno, Czech Republic*, CD-ROM (pp. 599–614).

Moellering, H. (2001b). *Research Needs in Analytical Cartography.* http://icaci.org/files/documents/ICCproceedings/ICC2001/icc2001/file/f24017.pdf.

Moellering, H. (2012). The scope of analytical cartography and its relationship to GIScience. *Proceedings AutoCarto 2012 & CaGIS 2012.*

Monk, R. (2005). *How to Read Wittgenstein.* London: Granta Book.

Monmonier, M. (1996). *How to Lie with Maps* (2nd ed.). Chicago: University of Chicago Press.

Montello, D. (2002). Cognitive map-design research in the twentieth century: Theoretical and empirical approaches. *Cartography and Geographic Information Science,29*(3), 238–304.

Moritz, H. (1995). *Science, Mind and the Universe: An Introduction to Natural Philosophy.* Heidelberg: Wichmann.

Morris, C. (1938). *Foundation of the theory of signs.* Chicago: Chicago University Press.

Morrison, J. (1986). The revolution in cartography in the 1980s. In D. W. Rhind & D. R. Fraser Taylor (Eds.), *Cartography: Past, Present and Future. A Festschrift for F.J. Ormeling.* London: Elsevier.

Murzi, M. (2004). Vienna Circle. Internet Encyclopedia of Philosophy IEP. A Peer-Reviewed Academic Resource. Retrieved December, 2010, http://www.iep.utm.edu/v/viennacircle.htm.

Nesher, D. (1997). Peircean realism: Truth as the meaning of cognitive signs representing external reality. *Transactions of the Charles S. Pierce Society, 33*(1), 201.

Neytchev, P. (2001). Concerning syntactic components of cartographic sentence. A. Wolodtschenko (ed.), *The Selected Problems of Theoretical Cartography 2000* (pp. 31–39) Proceeding of a Seminar of the Commission of Theoretical Cartography (Dresden, Germany, 14–14 October, 2000).

Nöth, W. (2005). Zeichentheoretische Grundlagen der Bildwissenschaft. In: Sachs-Hombach, K. (Ed.): *Bildwissenschaft zwischen Reflexion und Anwendung* (pp. 33–44). Cologne: Halem.

Nyerges, T. (1980). *Modelling the Structure of Cartographic Information for Query Processing*. Unpublished Ph.D. Dissertation, Ohio State University, Columbus, Ohio, 203 pp.

Ogrissek, R. (1982). *Erkenntnistheoretische Grundlagen und Erkenntnisgewinn in der Kartographie. Studienhilfsmittel.Lehrgebiet: Theoretische Kartographie*. Kartographische Bausteine, *3*, 78 p. Dresden: Institute for Cartography of University of Technology.

Ogrissek, R. (1987). *Theoretische Kartographie*. VEB Hermann Haack, Geographish-Kartographische Anstalt, Gotha.

Ormeling, F. (1996). Functionality of electronic school atlases. In B. Köbben, F. Ormeling & T. Trainor (Eds.), *Seminar on Electronic Atlases II*. ICA Proceedings on National and Regional Atlases (pp. 33–39). Prague: ICA.

Ormeling, F. (2007). Core concepts in cartographic communication. In E. Heere & M. Storms (Eds.), *Ormeling's Cartography:Presented to Ferjan Ormeling on the occasion of his 65th birthday and his retirement as Professor of Cartography* (pp. 73-80). Netherland Geographical Studies, 365. Faculteit Geowetenschappen Universiteit Utrecht.

Ormeling, F. (2010). From Ortelius to OpenStreetMap—Transformation of the map into a multifunctional signpost. In G. Gartner & F. Ortag (Eds.), *Cartography in Central and Eastern Europe* (pp. 1–16). Berlin: Springer.

Ormeling, F., & Kraak, M. (1987). *Kartographie: Ontwerp, productie en gebruik van kaarten* (p. viii). Delft: Delft University Press.

Ortega Valcárcel, J. (2000). *Los Horizontes de la Geografía: Teoría de la Geografía*. Barcelona: Editorial Ariel S.A.

Ostrowski, W. (2008). Stages of development of cartography as a science. *Miscellanea Geographica,3*, 267–276.

Padrón, R. (2004). *The Spacious World. Cartography, Literature, and Empire in the Early Modern Spain*. Chicago: University of Chicago Press.

Pajares, F. (2005). *The Structures of Scientific Revolution*. Outline and Study Guide, Philosopher's Web Magazine, Emery University Home Page. Retrieved January, 2011, from http://www.des.emory.edu/mfp/kuhn.html.

Pápay, G. (2005a). Die Beziehung von Kartographie, allgemeiner Bildwissenschaft und Semiotik. In: Sachs-Hombach, K. (Ed.): *Bildwissenschaft zwischen Reflexion und Anwendung* (pp. 33–44). Cologne: Halem.

Pápay, G. (2005b). Kartographie. In K. Sachs-Hombach (Ed.), *Bildwissenschaft zwischen Reflexion und Anwendung* (pp. 281–295). Cologne: Halem.

Pápay, G. (2005c). Bilder der Wissenschaft—Wissenschaft der Bilder: Einleitende Bemerkungen zu allgemeinen Bildwissenschaft. In G. Pápay (Ed.), *Bilder der Wissenschaft—Wissenschaft der Bilder* (pp. 7–18). Rostock: Philosophische Fakultät der Universität.

Pápay, G. (2006). Politics and cartography. In D. Unverhau (Ed.), *Sate Security and Mapping in the GDR: Map Falsification as a Consequence of Excessive Secrecy?* (pp. 1–13). Berlin: Lit Verlag.

Pápay, G. (2009). Kartographie. In S. Günzel (Ed.), *Raumwissenschaften* (pp. 158–173). Frankfurt am Main: Suhrkamp.

Pápay, G. (2012). Gedanken zur Kartographie als Bild- und Raumwissenschaft. Thoughts about Cartography as Image and Spatial Science. *Kartographische Nachrichten*, 62(1): 3–9.

Pape, H. (1989). *Charles Pierce zur Einführung*. Hamburg: Junius.

Papp-Váry, A. (1989). The Science of cartography. In D. Rhind & D. Taylor (Eds.), *Cartography: Past, Present and Future. A Festschrift for F.J. Ormeling* (pp. 103–109). London: Elsevier.

Paraskevopoulou, O., Charitos, D., & Rizopoulos, C. (2008). Location-Specific art practices the challenge the traditional conception of mapping. *Locative Media and Artistic Practice: Explorations on the Ground.* Artnodes. Iss. 8. UOC. Retrieved August, 2010, from http://www.uoc.edu/artnodes/8/dt/eng/paraskevopoupou_charitos_rizopoulos.pdf.

Perkins, C. (2003). Cartography: Mapping theory. *Progress in Human Geography,27*(3), 341–351.
Perkins, C. (2004). Cartography—Cultures of mapping: Power in practice. *Progress in Human Geography,28*(3), 381–391.
Perkins, C. (2007). Community mapping. *The Cartographic Journal,44*(2), 127–137.
Perkins, C. (2008). Cultures of map use. *The Cartographic Journal,45*(2), 150–158.
Peterson, M. (2002). The development of research in maps and the internet. In G. Gartner (Ed.), *Maps and the Internet* (pp. 1–6). Vienna: Technische Universität Wien.
Pickles, J. (2004). *A History of Spaces: Cartographic Reason, Mapping and the Geo-coded World*. London: Routledge.
Picker, M. (2013). Die Zukunft der Kartographie. Neue und nicht so neue epistemologische Krisen. In m. Picker, V. Maleval & F. Gabaude (Eds.), *Die Zukunft der Kartographie. Neue und nicht so neue epistemologische Krisen*. [NB: identical titles] (pp. 7–19). Bielefeld: transcipt Verlag.
Picker, M., Maleval, V. & Gabaude, F. (Eds. 2013). *Die Zukunft der Kartographie. Neue und nicht so neue epistemologische Krisen* (283 pp). Bielefeld: transcipt Verlag.
Pierce, Ch. S. (1978). *Écrits sur le ligne*. (Compiled, translated, commented by Gérard Delledalle). Paris: Seuil.
Popper, K., & Eccles, J. (1993). *The Self and its Brain: An Argument for Interactionism*. London: Routledge.
Ramirez, R. (2004). *Theoretical Cartography*. Book draft. http://www.cfm.ohio-state.edu/raul/document/.
Ramirez, R. (2001). New geographic visualization tool: A Multiple Source, quality, and media (MSQM) maps. http://icaci.org/files/documents/ICCproceedings/icc2001/file/f16019.pdf.
Ratzel, F. (1896). *The History of Mankind*. London: MacMillan and Co Ltd.
Ratzel, F. (1885). *Anthropogeographie öder Grundzüge der Erdkunde auf die Geschichte*. J. Engelhorn, Stuttgart,1 vol.
Reclus, E. (1876–1894). *Nouvelle Géographie Universelle*. Hachette, Paris, 19 vols.
Reichenbach, H. (1957). *The Philosophy of Space and Time*. New York: Dover.
Reyes, M. del Carmen (2005). Cybercartography from a modeling perspective. In D.R. Fraser Taylor (Ed.), *Cybercartography: Theory and Practice* (pp. 63–98). Amsterdam: Elsevier.
Reynolds, J. (2002). Jacques Derrida. Internet Encyclopedia of Philosophy IEP. A Peer-Reviewed Academic Resource. Retrieved January, 2011, from http://www.iep.utm.edu/d/derrida.htm.
Richter, D. (2004a). Ludwig Wittgenstein (1889–1951). Internet Encyclopedia of Philosophy IEP, A Peer-Reviewed Academic Resource. Retrieved November, 2010, from http://www.iep.utm.edu/w/wittgenstein.htm.
Richter, D. (2004b). *Historical Dictionary of Wittgenstein's Philosophy*. Historical Dictionaries of Religions, Philosophies, and Movements. Lanham: The Scarecrow Press.
Robinson, A. (1952). *The Look of Maps: An Examination of Cartographic Design*. Madison: University of Wisconsin Press.
Robinson, A., Sale, R., & Morrison, J. (1978). *Elements of Cartography* (4th ed.). New York, Wiley.
Robinson, A., Morrison, J., Muehrcke, P., Kimerling, J., & Guptill, S. (1995). *Elements of Cartography* (6th Ed.). New York: Wiley.
Rogoff, I. (2000). *Terra infirma. Geography's Visual Culture*. London: Routledge.
Romaniuk, S., & Paillalef, A. (2010). *Thomas Kuhn*. Retrieved November, 2010, from http://www.hechohistorico.com.ar/Archivos/Filosofia/Unidad4-Kuhn.pdf.
Russel, B. (1948). *Human Knowledge: Its Scope and Limits*. London: George Allen & Unwin.
Sachs-Hombach, K. & Schirra, J. R. J. (2000). Von der Bildwissenschaft zur Computervisualistik. In: J. Bernhard & G. Withalm (Eds.), Ikonische Zeichen. Auswahlakten zweier Konferenzen. *Semiotische Berichte 2000*, *24*(1–4): 287–304.

Saldivia, Z. (2010). Principales paradigmas que han marcado la evolución de las ciencias. Per Laboris ad Astram. Retrieved November, 2010, from www.perlaborisadastram-zenobio.blogspot.com/.

Stallmann, T. (2013). Living cartography. Book review: Atlas of Design. Volume 1. R. Wallace & D. P. Huffman (Eds.), XI + 81 p.) North American Cartographic Information Society, 2012. American Scientist, January–February 2013: 2. p.http://www.americanscientist.org/bookshelf/pub/living-cartography.

Sanchez, C. (2012). StereoWebMap: A Real Stereoscopic Map Server for the Internet. In M. F. Buchroithner (Ed.), *True-3D in Cartography—Autostereoscopic and Solid Visualisation of Geodata* (pp. 435–446)., Springer Lecture Notes in Geoinformation and Cartography Berlin: Springer.

Schiewe, J. (2013). Geovisualisation and Geovisaul Analytics: The Interdisciplinary Perspective on Cartography. Geovisualisierung und Geovisuelle Analytik: Die interdisziplinäre Sicht auf die Kartographie. *Kartographische Nachrichten*, 63, Special ICC 2013 Issue, 122–126.

Schiewe, J. & Schweer, M. (2013). Vertrauen im Rahmen der Nutzung von Karten. Trust in the Field of Map Usage. *Kartographische Nachrichten, 2/3,* 59–66.

Schlichtmann, H. (1985). Characteristic traits of the semiotic system 'map symbolism'. *The Cartographic Journal,22*(1), 23–30.

Schlichtmann, H. (1999). Map semiotics around the world. In H. Schlichtmann (Ed.), *Proceedings of a Seminar of the Working Group on Map Semiotics.* Dresden, Germany, June 20–21, 1997. International Cartographic Association.

Schlichtmann, H. (2009). Overview of the semiotics of maps. *Proceedings,24th International Cartographic Conference, Santiago de Chile, 15–21 November 2009.* Cd, 12 pp.

Schmithüsen, J. (1970). *Geschichte der Geographischen Wissenschaft: Von den Ersten Anfängen bis zum Ende des 18. Jahrhunderts.* Bibliographisches Institut, Mannheim/Wien/Zürich: Hochschultaschenbücher-Verlag.

Schmundt, H. (2011). Kriegsschiffe auf der Wiese. Vom Handy bis zum Geldautomaten – überall sind GPS-Chips versteckt. Jetzt warnen Experten: Wenn Störsender die Navigationssignale verwirren, spielt die Technik verrückt. *Spiegel*, 2011/26, 128–129.

Schulte, J. (2005). Ludwig Wittgenstein. Leben. Werk. Wirkung. Frankfurt/Main: Suhrkamp Publishers

Schwiersch, M. (2011). denn sie wissen nicht, wo sie sind. *bergundsteigen, 19*(1), 28–31.

Shannon, C. (1948). A mathematical theory of communication. *The Bell System Technical Journal,27*(379–423), 634–56.

Sheppard, E. (2005). Knowledge though critical GIS: Genealogy and prospects. *Cartographica,40*(4), 5–21.

Slife, B., & Williams, R. (1995). *What's Behind the Research?: Discovering Hidden Assumptions in the Behavioral Sciences.* Thousand Oaks: Sage Publications.

Slocum, T., McMaster, R., Kessler, F., & Howard, H. (2007). *Thematic Cartography and Visualization* (3rd ed.)., Series in Geographic Information Science Upper Saddle River: Prentice Hall.

Stark, W., & Brandt, R. (2009). Immanuel Kant: Vorlesungen über physische Geographie. Immanuel Kant; Gesammelte Schriften 26,1.- LXXX, 374 S.

Stockhammer, R. (2007). *Kartierung der Erde. Macht und Lust in Karten und Literatur*. Munich: Fink.

Sui, D., & Holt, J. (2008). Visualizing and analyzing public-health data using value-by-area cartograms: toward a new synthetic framework. *Cartographica,43*(1), 3–20.

Taylor, D. (1994). Cartography for Knowledge, action and development: Retrospective and prospective. *The Cartographic Journal,31*(1), 52–55.

Taylor, D. (2005). The theory and practice of cybercartogarphy: An introduction. In D. Taylor (Ed.) *Cybercartography: Theory and Practice* (pp. 1–14). Amsterdam: Elsevier.

Taylor, D. (2009). Some New applications in the theory and practice of cybercartography: Mapping with indigenous people in Canada's north. Retrieved October, 2010, from http://www.icaci.org/documents/ICC_proceedings/ICC2009/html/refer/22_1.pdf.

Thomas, J., & Cook, K. (2005). *Illuminating the Path: Research and Development Agenda for Visual Analytics*. Richland: IEEE Press.

Thrift, N. (2000). Non-representational theory. In R. J. Johnson, D. Gregory, G. Pratt, & M. Watts (Eds.), *The Dictionary of Human Geography*. Oxford: Blackwell.

Tobler, W. (1976). Analytical Cartography. *The American Cartographer,3*(1), 21–31.

Tobler, W. (1979). A transformational view of cartography. *American Cartographer,6*(2), 101–106.

Turnbull, D. (2003). *Masons, Tricksters and Cartographers: Comparative Studies in the Sociology of Scientific and Indigenous Knowledge*. London: Routledge.

Uebel, T. (2006). Vienna Circle. The Stanford Encyclopedia of Philosophy (Winter 2010 Edition) Edward N. Zalta (Ed.). Retrieved November, 2010, from http://plato.stanford.edu/archives/win2010/entries/vienna-circle/.

United Nations Department of Social Affairs (1949). Modern Cartography. Base Maps for World Needs. Code 32799 – Dec. 1949–3700. Sales No. 1949. I. 19. 95 p. Lake Success, New York 1949.

Unwin, D. (1994). Cartography ViSC and GIS. *Progress in Human Geography,18*(4), 516–522.

Vanecek, E. (1980). *Experimentelle Beitrage zur Wahrnehmarbeit Kartographischen Signaturen* (146 p). Wien. Publishing House of Austrian Acad. Science.

Vargas-Mendoza, J.E. (2006). Teoría del Conocimiento. México: Asociación Oaxaqueña de Psicología A.G. Retrieved November, 2010, from http://www.conductitlan.net/conocimiento.ppt.

Vico, E., & Bentancor, N. (1999). Evolución y Método de la Geografía. ANEP Departamento de Educación a Distancia. Edición 1999.

Wallace, T. R., & Huffman, D. P. (Eds.). (2012). *Atlas of Design* (Vol. 1)., North American Cartographic Information Society Milwaukee: Wisconsin.

Weiss, G. & Buchroithner, M. F. (2012). How True is True-3D? - Geometry between Single Image Stereograms and Lenticular Foil Displays. In M. F. Buchroithner (Ed.), *True-3D in Cartography—Autostereoscopic and Solid Visualisation of Geodata* (pp. 23–41), Springer Lecture Notes in Geoinformation and Cartography Berlin: Springer.

Wittgenstein, L. (1921). Logisch-philosophische Abhandlung, W. Ostwald (Ed.): *Annalen der Naturphilosophie, 14,* 185–262.

Wittgenstein, L. (1922). *Tractatus logico-philosophicus.* First published in English in 1922 by Routledge & Kegan Paul Ltd. Translated by C. K. Ogden, Reprinted 2005 by Routledge: New York.

Wolodtschenko, A. (2001). Kartosemiotische Systeme und Konzeptionen in Osteuropa. In W.G. Koch (Ed.), *Theorie 2000* (pp. 89–94). Kartographische Bausteine, Vol. 19, Technische Universität Dresden. Dresden 2001.

Wolodtschenko, A. (2003). Structural Models in Cartography. In A. Wolodtschenko (Ed.), *The Selected Problems of Theoretical Cartography 2002* (pp. 73–81). Dresden: Institute for Cartography, Dresden University of Technology.

Wolodtschenko, A. (2008). Über die konzeptionelle Bewegung in der Kartographie. *meta-carto-semiotics,Journal for Theoretical Cartography1*, 1–16. http://meta-carto-semiotics.org.

Wolodtschenko, A. (2009). Einige Erinnerungen an die Zukunft der Kartographie. In A. Wolodtschenko & H. Schlichtmann (Eds.), *Diskussionsbeiträge zur Kartosemiotik und zur Theorie der Kartographie 12,* 43–60. Dresden: Selbstverlag der Technischen Universität Dresden.

Wolodtschenko, A. (2011). Quo Vadis Classic Cartosemiotics & Quo Vadis Theoretical Cartography? *meta-carto-semiotics,Journal for Theoretical Cartography* (4), 1–18. http://meta-carto-semiotics.org

Wolodtschenko, A. (2012). On the new generation of digital mini-atlases. *Proceedings AutoCarto 2012 & CaGIS 2012.*

Wood, D. (1992). *The Power of Maps*. New York: Guilford Press.

Wood, D., & Fels, J. (2008). *The Natures of Maps: Cartographic Constructions of the Natural World*. Chicago: University of Chicago Press.

Wood, D., & Fels, J. (2008). The nature of maps: Cartographic constructions of the natural world. *Cartographica,43*(3), 189–202.

Wood, D., & Krygier, J. (2009). Critical cartography. In N. Thrift & R. Kitchen (Eds.), *The International Encyclopledia of Human Geography*. New York: Elsevier.

Woodward, A. (2005). Jean-Francois Lyotard (1924–1998). Internet Encyclopedia of Philosophy IEP A Peer-Reviewed Academic Resource. http://www.iep.utm.edu/l/lyotardl.htm.

Wright, G. (1971). *Explanation and Understanding*. Ithaca, New York: Cornell University Press. Translated Spanisch *Explicación y Comprensión*, Madrid: Alianza Editorial, 1979.

Zarycki, T. (2001a). On the pragmatic approach to map analysis: Remarks on the basis of MacEachren's approach to map semiotics. In A. Wolodtschenko (Ed.) *The Selected Problems of Theoretical Cartography 2000* (pp. 64–70). Proceedings of a Seminar of Commission on Theoretical Cartography, Institute for Cartography, Dresden University of Technology, Dresden.

Zarycki, T. (2001b). Cartographic Communication in the Perspective of the Linguistic Pragmatics. Scientific and Technical Program Committee LOC for ICC Beijing China (Ed.) *Proceedings 5 of the 20th Cartographic Conference ICC 2001 Beijing China Mapping the 21th Century* (pp. 3099–3108) Chinese Society of Geodesy Photogrammetry and Cartography.

Zentai, L. (2012). Does Cartography still exist? *Proceedings AutoCarto 2012 & CaGIS* 2012.

MIX
Papier aus verantwortungsvollen Quellen
Paper from responsible sources
FSC® C105338

Printed by Books on Demand, Germany